T0229497

MOBILE, WIRELESS AND SENSOR NETWORKS

A Clustering Algorithm for
Energy Efficiency and Safety

MOBILE, WIRELESS AND SENSOR NETWORKS

A Clustering Algorithm for Energy Efficiency and Safety

Amine Dahane, PhD
Nasr-Eddine Berrached, PhD

APPLE ACADEMIC PRESS

Apple Academic Press Inc.
3333 Mistwell Crescent
Oakville, ON L6L 0A2 Canada

Apple Academic Press Inc.
1265 Goldenrod Circle NE
Palm Bay, Florida 32905 USA

Library and Archives Canada Cataloguing in Publication

Title: Mobile, wireless and sensor networks : a clustering algorithm for energy efficiency and safety / Amine Dahane, Nasr-Eddine Berrached, PhD.

Names: Dahane, Amine, author. | Berrached, Nasr-Eddine, author.

Description: Includes bibliographical references and index.

Identifiers: Canadiana (print) 20189068442 | Canadiana (ebook) 20189068450 | ISBN 9781771886796 (hardcover) | ISBN 9781351190756 (e-book) | ISBN 9781351190749 (e-book ; PDF)

Subjects: LCSH: Wireless sensor networks—Mathematical models.

Classification: LCC TK7872.D48 D34 2018 | DDC 681.2—dc23 | 006.2/5—dc23

Library of Congress Cataloging-in-Publication Data

Names: Dahane, Amine, author. | Berrached, Nasr-Eddine, author.

Title: Mobile, wireless and sensor networks : a clustering algorithm for energy efficiency and safety / Amine Dahane. Nasr-Eddine Berrached, PhD.

Description: Toronto ; New Jersey : Apple Academic Press, 2019. | Includes bibliographical references and index.

Identifiers: LCCN 2018058119 (print) | LCCN 2018060218 (ebook) | ISBN 9781351190756 (ebook) | ISBN 9781771886796 (hardcover : alk. paper)

Subjects: LCSH: Wireless communication systems. | Mobile computing. | Cluster analysis. | Computer algorithms. | Cooperating objects (Computer systems)

Classification: LCC TK5103.2 (ebook) | LCC TK5103.2 .D314 2019 (print) | DDC 004.6028/6--dc23

LC record available at https://lccn.loc.gov/2018058119

Apple Academic Press also publishes its books in a variety of electronic formats. Some content that appears in print may not be available in electronic format. For information about Apple Academic Press products, visit our website at **www.appleacademicpress.com** and the CRC Press website at **www.crcpress.com**

DEDICATION

- To my parents
- My brothers and sisters
- My grandparents
- My colleagues at LARESI and all my friends, especially, Abdlhak Bentaleb, Dr. Messaoud Bensaada, Dr. Idriss Messaoudene, Freih Bengabbou Mokhtar, and my best friend, Mohammed Yahyaoui.
- To Professor Khelfaoui Rachid (University of Bechar, Algeria)
- To Professor Derdour Aicha (University of USTO, Algeria)
- To Professor Maroc Mohammed (University of Tlemcen, Algeria)
- To Professor Shakshuki Elhadi (University of Acadia, Canada)

ABOUT THE AUTHORS

Amine Dahane, PhD, is affiliated with the Research Laboratory in Industrial Computing and Networks (RIIR), University of Oran 1 Ahmed Benbella, Oran 31000, Algeria. His main research interests include wireless sensor networks, their security, routing and management, intrusion detection, and MAC protocols design issues. He is a reviewer of several journals and participates regularly at professional conferences. He received his master's degree in Computer Systems and Networks from the University of Bechar and his PhD degree in Electronics from the University of Sciences and Technology of Oran (USTO, Algeria). He is currently a lecturer and researcher at the institute of Sciences and Applied Techniques of the University of Oran 1, Ahmed Benbella (Algeria) where he teaches Instrumentation, Informatics, networking, Programming languages.

Nasr-Eddine Berrached, PhD, is a professor in the Electronics Department at University of Science and Technology of Oran, Algeria. Previously he headed the laboratory of robotics and the research laboratory in intelligent systems (LARESI). His interests include man-machine interface, telerobotics, machine vision, pattern recognition and inverse problem. He received his doctor of engineering in computer science degree from the Tokyo Institute of Technology (TIT, Japan).

CONTENTS

AUTHOR CONTACT INFORMATION

AMINE DAHANE
Research Laboratory in Industrial Computing and Networks (RIIR),
University of Oran 1 Ahmed Benbella, Oran 31000, Algeria,
E-mail: amine.dahane@univ-oran.dz

NASR-EDDINE BERRACHED
Electronics Department, University of Science and Technology of Oran,
Bir El Djir, Algeria, E-mail: laresi.usto.2015@gmail.com

ACRONYMS

ACK	acknowledgment message
AODV	ad hoc on demand distance vector
ARQ	automatic repeat request
BL_i	behavior level of node
BS	base station
CH	cluster head
C_i	connectivity
CMs	cluster members
DHCEA	dynamic head clustering election algorithm
D_i	distance between node and its neighbors
DOS	denial of service
DSDV	destination sequenced distance vector
DWCA	distributed weighted clustering algorithm
Er_i	energy residual
ES-WCA	energy-efficient and safe weighted clustering algorithm
FEC	forward error correction
G	gateway
LAN	local area network
LE	leader election message
MAC	medium access control
MACs	message authentication codes
MAN	medium area network
MANET	mobile ad hoc network
MaSE	multi-agent system engineering
MEMS	micro-electromechanical systems
M_i	mobility
MT	mobile terminal
ND	neighbor discovering message
NMEICT	National Mission on Education through ICT
OMNet++	Objective Modular Network Test Bed in C++
ONL	open network laboratory
PARSEC	PARallel simulation environment for complex
PCH	primary cluster head

PKC	public key cryptography
PW	practical work
RECA	reputation-based clustering algorithm
RERR	route error message
RLI	remote laboratory interface
RREP	route reply message
RREQ	route request message
SCH	secondary cluster head
SKE	systematical key encryption
SMP	sensor management protocol
SPEED	A Stateless Protocol for Real-Time Communication in Sensor Networks
SQDDP	sensor query and data dissemination protocol
TADAP	task assignment and data advertisement protocol
Tcl	tool command language
TCP	transmission control protocol
TTL	time to live
UDP	user datagram protocol
UML	unified modeling language
VLP	virtual laboratory platform
VM	verifiable multilateration
WCA	weighted clustering algorithm
WSNs	wireless sensor networks

PREFACE

Wireless networking is a research field that attracts more and more attention among researchers. It includes a variety of topics involving many challenges. The main concern of clustering approaches for mobile wireless sensor networks (WSNs) is to prolong the battery life of the individual sensors and the network lifetime. For a successful clustering approach, the development of a powerful mechanism to safely elect a cluster head remains a challenging task in many research works that take into account the mobility of the network. The approach based on the computing of the weight of each node in the network is one of the proposed techniques to deal with this problem.

In this book, we propose a distributed and safe weighted clustering algorithm (ES-WCA) for mobile WSNs using a combination of five metrics. Among these metrics lies the behavioral-level metric that promotes a safe choice of a cluster head in the sense where this last one will never be a malicious node. Moreover, the highlight of our work is summarized in a comprehensive strategy for monitoring the network in order to detect and remove the malicious nodes. We use simulation study to demonstrate the performance of the proposed algorithm. Furthermore, we present a virtual laboratory platform of baptized Mercury, allowing students to make practical work on different aspects of mobile WSNs. Our choice of WSNs is motivated mainly by the use of real experiments needed in most courses about WSNs. These experiments require an expensive investment and a lot of notes in the classroom. The platform presented here aims to show the feasibility, the flexibility, and the reduced cost of such a realization. We demonstrate the performance of the proposed algorithms that contribute to the familiarization of the learners in the field of WSNs.

ACKNOWLEDGMENTS

This work would not have been possible without the support of many people I want to thank. First of all, I thank Allah who helped me to develop this book that is a revised version of my PhD thesis.

No words can express my gratitude to my parents, Yasmina Khelfaoui and Mohammed Dahane. I am forever grateful to them for their love, support, prayers, encouragement, sacrifices, and help throughout my life.

I would like to thank Professors Nasr-Eddine Berrached, my supervisor, Abdelhamid Loukil, and Bouabdellah Kechar for their invaluable support, guidance, and encouragement throughout these years. I feel very privileged to have had the opportunity to learn from and work with them. I hope I managed to adopt some of their professional attitude and integrity. I express my sincere and deepest gratitude to my examination committee member, Professor Ouameri, for accepting to be the chairman of the examination committee and the other members of my committee, Professor Ouslim, Professor Hendel, Professor Zerhouni, and Professor Cherki, for their kind willingness to judge this work.

On a personal note, I would like to extend my heartfelt gratitude to all the members for our amazing laboratory, LARESI, for their respect and support.

I would also like to thank my best friend, Mohammed Yahyaoui, who helped me to find a studio near Croydon (South London), as I have a conference in London that is very important for my scope of research and that helped me to complete my PhD thesis.

Further, I would like to thank Abdlhak Bentaleb, Chirihane Gherbi, Nassima Merrabtine, Amine Cheriet, and Aness Belhouari for meetings and discussions in the sensor network field.

At the last but not the least, I would like to thank my parents, my brother, and my sisters for unquestioningly supporting and encouraging me, for a long as I can remember, in whatever I chose to do.

Finally and with great emotion, I thank my family, my teachers of primary, middle, secondary, university, and all my relatives for making all this possible.

—**Amine Dahane**

CHAPTER 1

WIRELESS SENSOR NETWORKS: A SURVEY

1.1 INTRODUCTION

Recent advances in micro-electromechanical systems in wireless communications and in digital electronics made possible the production of small, cheap, and "smart" devices such as personal digital assistants (PDAs), radio frequency identification systems, wireless sensor networks (WSNs), and many other technologies. In this chapter, we focus on the security issues of the representative technology of WSNs.

A sensor device is a small device that is able to sense environmental data. These tiny sensor nodes, which consist of sensing, data processing, and communicating components, leverage the idea of sensor networks based on collaborative effort of a large number of nodes. Sensor networks represent a significant improvement over traditional sensors and are deployed in the following two ways[6]:

- Sensors can be positioned far from the actual phenomenon, that is, something known by sense perception. In this approach, large sensors that use some complex techniques to distinguish the targets from environmental noise are required.
- Several sensors that perform only sensing can be deployed. The positions of the sensors and communications topology are carefully engineered. They transmit time series of the sensed phenomenon to the central nodes where computations are performed and data are fused.

A sensor network is composed of a large number of sensor nodes, which are densely deployed either inside the phenomenon or very close to it.

While many sensors connect to controllers and processing stations directly (e.g., using local area networks), an increasing number of sensors communicate the collected data wirelessly to a centralized processing station. This is important since many network applications require hundreds or thousands of sensor nodes often deployed in remote and inaccessible areas. Therefore, a wireless sensor has not only a sensing component but onboard processing, communication, and storage capabilities also. With these enhancements, a sensor node is often not only responsible for data collection but also for in-network analysis, correlation, and fusion of its own sensor data and data from other sensor nodes.

When many sensors cooperatively monitor large physical environments, they form a WSN.

Sensor nodes communicate with each other as well as with a base station (BS) using their wireless radios, which allow them to disseminate their sensor data to remote processing, visualization, analysis, and storage systems. For example, Figure 1.1 shows two sensor fields monitoring two different geographic regions and connecting to the internet using their BSs.

The capabilities of sensor nodes in a WSNs can vary widely, that is, simple sensor nodes may monitor a single physical phenomenon, while more complex devices may combine many different sensing techniques (e.g., acoustic, optical, and magnetic). They can also differ in their communication capabilities, for example, using ultrasound, infrared, or radio frequency technologies with varying data rates and latencies. While simple sensors may only collect and communicate information about the observed environment, more powerful devices (i.e., devices with large

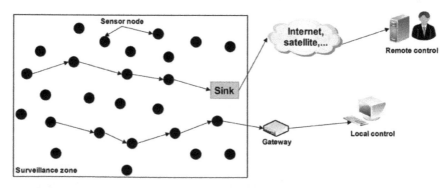

FIGURE 1.1   Example of wireless sensor networks.

processing, energy, and storage capacities) may also perform extensive processing and aggregation functions. Such devices often assume additional responsibilities in WSNs, such as they may form communication backbones that can be used by other resource-constrained sensor devices to reach the BS.

1.2 DEFINITION

WSNs are a special type of ad hoc networks defined by cooperative sensor nodes set dispersed in a geographic zone, called capture zone, in order to monitor a phenomenon and collect its data independently.

A sensor network is composed of a large number of sensor nodes that are deployed randomly in high density within the capture area or close to it, and having cooperative capabilities. One WSN consists of tens to hundreds and sometimes thousands of sensors nodes. These nodes are autonomous entities able to perform three complementary tasks: collect data, process them, and then communicate through the network (Fig. 1.2).

FIGURE 1.2   A sensor node.

The sensor nodes work in a cooperative manner by means of their sharing capabilities of collected information through the air. The choice of the wireless connection rather than the wired link allows great flexibility of use in addition to an easy and fast deployment. The sensors must also be equipped with a battery which is unsustainable power source and not replaceable in most cases—the most important constraint of network survivability.

Sensor networks may consist of many different types of sensors, such as seismic, low sampling rate magnetic, thermal, visual, infrared, acoustic, and radar, which are able to monitor a wide variety of ambient conditions that include the following:

- Temperature
- Humidity
- Vehicular movement

- Lightning condition
- Pressure
- Soil makeup
- Noise levels
- The presence or absence of certain kinds of objects
- Mechanical stress levels on attached objects
- The current characteristics such as speed, direction, and size of an object

Realization of these and other sensor network applications require wireless ad hoc networking techniques. Although many protocols and algorithms have been proposed for traditional wireless ad hoc networks, they are not well suited for the unique features and application require-ments of sensor networks. To illustrate this point, the differences between sensor networks and ad hoc networks are outlined below:

- The number of sensor nodes in a sensor network can be several orders of magnitude higher than the nodes in an ad hoc network.
- Sensor nodes are densely deployed.
- Sensor nodes are prone to failures.
- The topology of a sensor network changes very frequently.

1.3   ARCHITECTURE OF A WIRELESS SENSOR NETWORKS

WSNs are typically composed of a well field, several sensors nodes, a BS, and a data processing center (see Fig. 1.3). These are discussed in detail below.

- Well field (collection area): It is considered to be the area of interest for the sensed phenomenon, so the sensor nodes are placed here.
- Sensor nodes: These are the hearts of the network. Their role is to collect data and route them to the BS. Their energy is often limited since they are powered by batteries.
- Sink (BS): This is a particular node responsible for receiving, storing, and processing data from other nodes and disseminates various requests to the network. Its energy source should be unlimited because it must always remain active to receive the data.

The sink can be a single node with additional network interfaces (usually Ethernet or GPRS) or an internal entity to the network as a laptop or PDA.

- Data processing center (task manager): It receives data collected by the sink; its role is to group and process received data in order to extract the useful information.

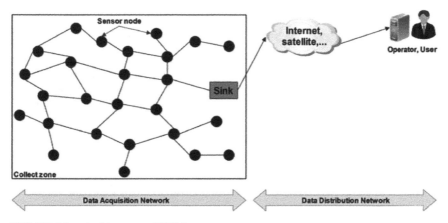

FIGURE 1.3   Architecture of WSNs.

As shown in Figure 1.3, WSNs consist of many sensor nodes scattered on the well field. When the sink broadcasts a request, nodes cooperate with each other to send it the information captured through a "multi-hop" architecture.

Thereafter, the sink transmits these data through internet or satellite task manager to analyze and make decisions. At a higher level, WSNs can be seen as a combination of following two network entities:

- The data acquisition system: It is the union of the sensor nodes and the sink. Its role is to collect data from the environment and gather them in sink.
- The data distribution network: Its role is to connect the network data acquisition to the user.

Architectures in sensor networks depend on the application and the techniques used to route the sensor information to the BS. Taxonomy of applications can be derived and adaptability algorithms for this kind of scenario can be assessed.

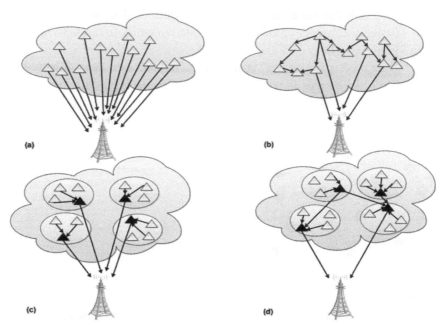

FIGURE 1.4   Data communication architecture in WSNs.
Source: Adapted from ref 11.

The routing process of the sensor information to the BS may take four forms. In flat architecture, the sensors may communicate directly with the BS using a high power (Fig. 1.4(a)) or via a multi-hop fashion with very low powers (Fig. 1.4(b)), and in hierarchical architectures, the cluster representative node, called cluster head, directly transmits data to the BS (Fig. 1.4(c)) or via a multi-hop fashion between cluster heads (Fig. 1.4(d)). There are two types of architecture for WSNs that are discussed in detail below.

1.3.1   FLAT WSNS

Flat WSN is a homogeneous network where all nodes have same capabilities and functions related to the capture, communication, and hardware complexity, only the sink is exception to this rule since it plays the role of a gateway responsible for the collection of data from different sensor nodes to transmit to the user (Fig. 1.5).

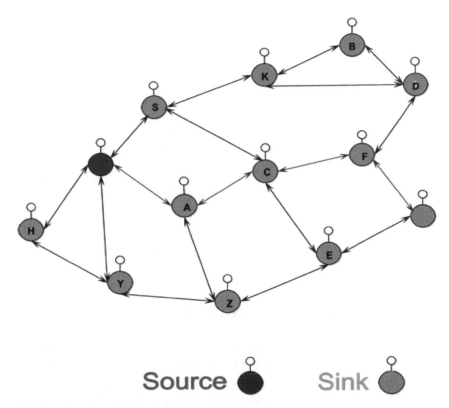

FIGURE 1.5 Flat architecture.

1.3.2 HIERARCHICAL WSNS

A network of sensors is reporting a heterogeneous network where nodes have different abilities, for example, certain nodes can have a more important source of energy, a longer communication range, and/or greater computing power. This allows discharging most simple nodes at the low cost of several functions of the network (Fig. 1.6).

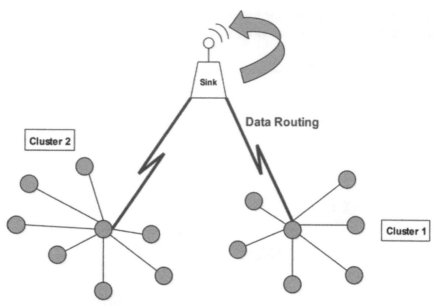

FIGURE 1.6   Hierarchical architecture.

1.4   DESIGN CONSTRAINTS OF WSNS

A WSN is a special network which has many constraints compared to a traditional computer network. Owing to these constraints, it is difficult to directly employ the existing security approaches to the area of WSNs. Therefore, in order to develop useful security mechanisms, while borrowing the ideas from the current security techniques, it is necessary to know and understand these constraints first.[3] The main factors influencing the stress and sensor network architecture can be summarized as follows[5]:

- Fault tolerance: Certain nodes may generate errors or stop working because of a lack of energy, a physical problem, or interference. These problems do not affect the rest of the network; it is the principle of fault tolerance. Fault tolerance is the ability to maintain network functionality without interruptions due to an error occurred on one or more sensors.
- Scale: The number of nodes deployed to a project may reach one million. An important number of nodes generates a lot of internodal

transmissions and requires that the well "sink" is equipped with a lot of memory to store the information received.

- Production costs: Often sensor networks are composed of a very large number of nodes. The price of a node is critical in order to compete with traditional monitoring network. Currently, nodes often only cost a little more than $US1. For comparison, a Bluetooth node, though already known to be a low-cost system, costs approximately $US10.

- Environment: The sensors are often deployed en masse in places such as battlefields beyond the enemy lines, inside large machines at the bottom of an ocean, in biologically or chemically contaminated areas, and so forth. Therefore, they must operate unattended in remote geographic areas.

- Network topology: The deployment of a large number of nodes requires maintenance of topology. This maintenance consists of three phases: deployment, post-deployment (the sensors can move, not work, etc.), and redeployment of additional nodes.

- Material constraints: The main constraint is the physical size of the sensor. The other constraints are that the energy consumption has to be reduced for the network to survive as long as possible; it adapts to different environments (high temperatures, water, etc.), whether autonomous and very resistant since it is often deployed in harsh environments.

- Transmission media: In WSNs, the nodes are connected by a wireless architecture. To enable operations on these networks around the world, the transmission medium must be standardized. Most often, infrared (which is license-free, robust to interference, and inexpensive), Bluetooth, and ZigBee radio communications are used.

- Energy consumption: A sensor because of its size is limited energy (< 1 W). In most cases, the replacement of the battery is impossible.

Dysfunction of some nodes requires a change in the network topology and rerouting of packets. All these operations are energy intensive; it is for this reason that current research focuses primarily on ways to reduce consumption.

1.5 SENSORS

In the following section, we will study in detail the components of a sensor node and some examples of existing nodes on the market.

1.5.1 PHYSICAL ARCHITECTURE OF A SENSOR NODE

Figure 1.7 shows a typical architecture of a sensor node. It contains four base units: the sensor unit, the processing unit, the transmission unit, and the power control unit. It may also contain additional modules that depend on applications such as global positioning system to know the precise location of the node, an energy generating system (solar cell), or even a mobilizing system responsible for the sensor movement if necessary.

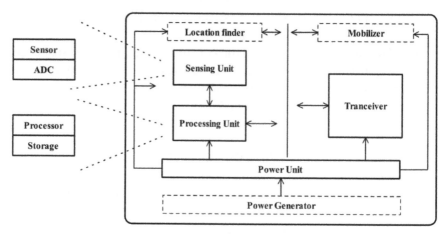

FIGURE 1.7   Logical components of a sensor unit.
Source: Adapted from ref 6.

In the subsequent text, we will detail the basic components of a sensor[1]:

- Capture unit: It captures observed phenomenon and converts it into a digital signal to be sent to the processing unit.
- Processing unit: This unit is responsible for managing the communication procedures that allow a node to collaborate with other

network nodes to perform the assigned tasks. It can also analyze the data collected to ease the task of the sink.

- Transmission unit: This unit manages the network connection node by performing all the transmission and reception of data over a wireless medium.
- Power control unit: It is one of the most important systems. It is responsible for distributing the available energy between the different units and reduces expenses, for example, pausing inactive components.

Figure 1.8 summarizes the energy consumption in the different units of a sensor node.

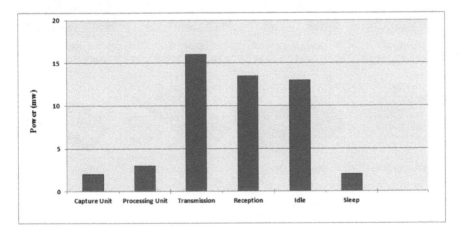

FIGURE 1.8   Energy consumption in a sensor node.
Source: Adapted from ref 2.

We notice that the transmission unit has very intensive energy consumption. Indeed, the transmission of a single bit along a distance of 100 m consumes approximately the same energy as the execution of 3000 instructions. However, the standby radio significantly reduces consumption.

1.5.2 TYPES OF SENSORS

There are currently a large number of sensors, with different and varied features. Most of these sensors depend on the application for which they were designed (water sensors, underground, etc.). Figure 1.9 shows the evolution of sensors over the last 20 years.

FIGURE 1.9   Sensors evolution.
Source: Adapted from ref 11.

The sensors manufactured by Crossbow over the past 10 years (sensor family Mica and Telos) are the most used in experiments and research. These sensors are able to measure several parameters (temperature, humidity, light, etc.) and most of them revolve around the Chipcon.

TABLE 1.1   Characteristics of the Most Used Sensors. (Source: Reprinted from ref 3 with permission.)

Sensor	MCU	RAM	Flash	Storage	Radio	Dimension
Spec Node (2003)	AVR Risk 8 bit	KB	KB	KB	RF	2*2.5 mm
Mica (2002)	ATMega 128	KB	KB	KB	CC1000	58*32*7 Mm
Mica2Dot (2002)	ATMega 128	KB	KB	KB	CC1000	25*6 Mm
MicaZ (2004)	ATMega 128	KB	KB	KB	CC2420	58*32*7 mm
TelosA (2004)	TI MSP 430	KB	KB	KB	CC2420	
TelosB (2004)	TI MSP 430	KB	KB	KB	CC2420	65*31*6 Mm
Tmote Sky (2004)	TI MSP 430	KB	KB	KB	CC2420	3.2*8*1.3 Mm
BT node3 (2004)	ATMega 128	KB	KB	KB	CC1000	58.15*33 Mm
Imote (2003)	ARM7	KB	KB	KB	ZV4200	
Imote2 (2007)	Intel PXA271	KB	KB	KB	CC2420	36*48*9 Mm
Iris (2008)	ATmega 1281	KB	KB	KB	CC2420	58*32*7 Mm

Table 1.1 summarizes the main characteristics of sensors of Xbow company (Crossbow)[7] and the most used sensors in the field of research.

1.6   APPLICATIONS OF SENSOR NETWORKS

WSNs have inspired many applications. Some of them are futuristic while a large number of them are practically useful. The diversity of applications in the latter category is remarkable—environment monitoring, target tracking, pipeline monitoring (water, oil, and gas), structural health monitoring, precision agriculture, health care, supply chain management, active volcano monitoring, transportation, human activity monitoring, and underground mining, to name a few. In the subsequent text, we present some useful applications of WSNs.

1.6.1 MILITARY APPLICATIONS

Some of the military applications are as follows:

- Monitoring friendly forces, equipment, and ammunition
- Recognition of opposing forces and terrain
- Battlefield surveillance
- Battle damage assessment
- Nuclear, biological, and chemical attack detection

The low cost, rapid deployment, self-organization, and fault tolerance are characteristics that have made effective sensor networks for military applications.[6] Several projects have been launched to help the military units in the battlefield and protect the city against attacks, such as terrorist threats (see Fig. 1.10).

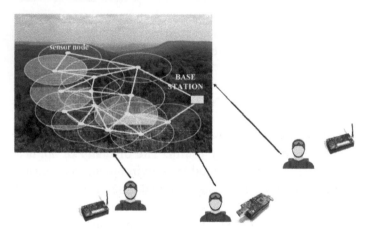

FIGURE 1.10   (See color insert.) Control of Armed Forces.
Source: Adapted from ref 11.

The Distributed Sensor Network project[10] in Defense Advanced Research Projects Agency was one of the first projects in the 1980s which used sensor networks to gather distributed data. Researchers at Lawrence Livermore National Laboratory have implemented network Wide Area

Tracking System.[12] This network consists of gamma rays and neutrons detectors to detect and screen for nuclear devices. It is capable of continuous monitoring of an area of interest. It uses data aggregation techniques to bring them to a smart center.

The researchers then set up another network called Joint Biological Remote Early Warning System[9] to warn the troops in the battlefield of possible biological attacks. A sensor network can be deployed in a strategic or hostile place, in order to monitor the movements of enemy forces or analyze the field before sending troops (chemical, biological, or radiation detection). The US military has conducted tests in the desert of California.

1.6.2 ENVIRONMENTAL APPLICATIONS

Another field of WSNs application is agriculture. Some of the applications are as follows:

- Flood detection
- Forest fire detection
- Environmental monitoring (see Fig. 1.11)
- Biocomplexity mapping of the environment
- Precision agriculture (see Fig. 1.12)

FIGURE 1.11   Environmental monitoring seabirds.
Source: Adapted from ref 11.

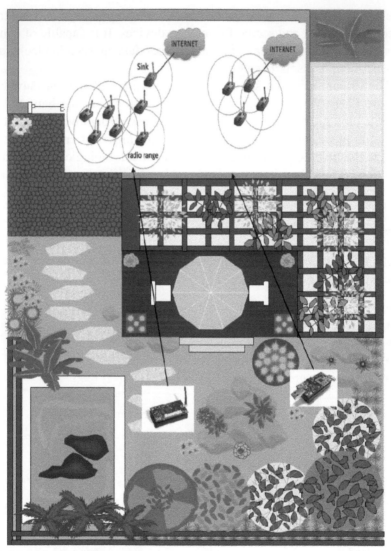

FIGURE 1.12   Precision agriculture.
Source: Adapted from ref 11.

1.6.3   HEALTH APPLICATIONS

Another field of high demand for such networks is medicine. We have cited some applications below:

- Tele-monitoring of human physiological data (see Fig. 1.13).
- Tracking and monitoring patients and doctors inside a hospital.
- Tele-monitoring of the elderly at home.
- Drug administration in hospitals.[4]

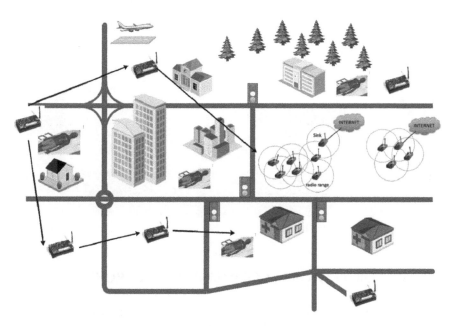

FIGURE 1.13 Remote monitoring of physiological information.
Source: Adapted from ref 11.

1.6.4 HOME AND OTHER COMMERCIAL APPLICATIONS

Some of the home and commercial applications are as follows:

- Home automation and smart environment (see Fig. 1.14)
- Interactive museums
- Managing inventory control
- Vehicle tracking and detection
- Detecting and monitoring car thefts

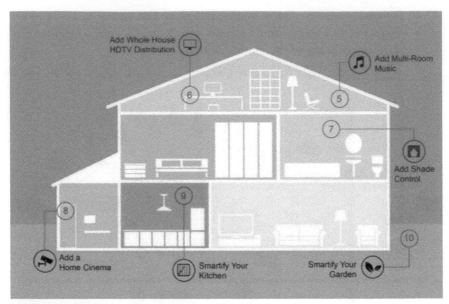

FIGURE 1.14   Automated home.

1.7   SENSOR NETWORKS COMMUNICATION ARCHITECTURE

The protocol stack used by sink and other sensor nodes is presented in Figure 1.15. This protocol stack supports the energy consumption problem, integrates the processing of data transmitted in the routing protocols, and facilitates cooperative work between the sensors. It is composed of five layers: application, transport, network, data link, and physical layer, in addition of three levels (shots) which are transversed. These are as follows:

- The energy management level: Responsible for controlling how a node uses its energy.
- The task management level: Ensures the balancing of tasks distribution on the various nodes to perform a cooperative work.
- The mobility management level: Detects and records all movements of sensor nodes.

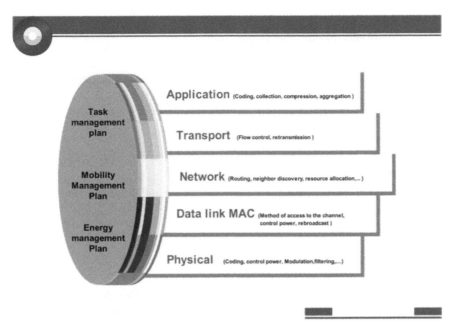

FIGURE 1.15 The protocol stack.

Now we explain the role of each of the five levels shown in the above figure.

1.7.1 PHYSICAL LAYER

This layer is responsible for addressing the needs of simple and robust modulation, transmission, and receiving techniques. This includes:

- The frequency selection
- The frequency carrier generation
- The signal detection and propagation
- The signal modulation and data encryption

1.7.2 DATA LINK LAYER

The data link layer is responsible for the multiplexing of data stream, data frame detection, the medium access (MAC), and error control.

- MAC
- Power-saving modes of operation
- Error control

1.7.2.1   MEDIUM ACCESS CONTROL (MAC)

- Creation of the network infrastructure
- Fairly and efficiently share communication resources between sensor nodes
- Existing MAC protocols (cellular system, Bluetooth, and mobile ad hoc network)

1.7.2.2   MAC FOR SENSOR NETWORKS

Table 1.2 summarizes the proposed taxonomy by Doudou et al.[8] of the protocols described throughout the chapter. The state-of-art literature presents a huge number of synchronous energy-efficient MAC protocols, which are all based on establishing common active/sleep schedules. However, each protocol has its own specific mechanism that deals with this trade-off. The concept of WSN MAC protocols timeliness has been investigated in this chapter, where a novel taxonomy has been introduced. Compared to the traditional classification based on medium access principles, the new classification is based on the solutions mechanisms that affect the transmission delay. Well-known energy-efficient synchronous contention-based MAC protocols have been discussed from the latency point of view. Two main categories from the literature based on the use of duty cycling were distinguished: static and adaptive duty-cycled protocols.

The latter has been divided into four different subclasses:

1.   Adaptive grouped schedule protocols
2.   Adaptive repeated schedule protocols
3.   Staggered schedule protocols
4.   Reservation schedule protocols

Table 1.2 summarizes the proposed taxonomy of the protocols described throughout the chapter. Our review indicates that some of these protocols can ensure end-to-end delay decrease, but none of them can provide delay guarantee for time-critical applications thus far. Consequently, huge efforts needed in order to enable successful deployment of WSNs for delay-sensitive applications.

TABLE 1.2 Summary of the Studied Slotted Contention-Based MAC Protocols.

Category	Protocols	Delay decrease	Latency	Cross layer support	Mechanism used affecting delay	Traffic/topology pattern
Static scheduled protocols	S-MAC (Ye et al. 2002)	1-Hop delay decrease	High due to fixed duty cycle	No	Fixed duty-cycle (long active period)	Bidirectional, random topology, unicast, broadcast
Adaptive grouped schedule	T-MAC (Van Dam and Langendoen, 2003), AD-MAC (Kim et al, 2008), MR-MAC (Zhao and Sun, 2008), LLM (Parmar et al, 2006), nanoMAC (Ansari et al, 2007), Adaptive Listenin (Ye et al, 2004), MRPM (Sthapit and Pyun, 2011), DSMAC (Lin et al, 2007), MD-MAC (Hameed et al, 2009), and FPA (Li et al, 2005)	2-Hops (T-MAC), K-hops (MAC) delay decrease where K depend on th duration of TA	Increase due to early sleeping and collisions	Routing	EAdaptive listening with incoming data notification message (FRT PRTS, \and LTR and time-out (TA))	Bidirectional, random topology, unicast, broadcast
Adaptive repeated schedule	SCP-MAC (Ye et al, 2006)	K-hops deay decrease where K is a number of active periods	Increases due to multi-hop latency (nodes can send only one packet per slot)	No	Using wakeup preamble, repeated small active period	Bidirectional, random topology, unicast, broadcast
Adaptive staggered schedule	D-MAC (Ye et al, 2002), LEMR (Cortes et al, 2009), L-MAC (Wang and liu, 2007), and Q-MAC (Vasanthi and Annadurai, 2006)	End-to-end delay devrease	Low latency for a very low data rate, high latency due to per-level collisions in dence network	Routing	Per-level cascading wakeup time	Tree (Unidirectional, Converge-cast)
Adaptive reservation schedule	R-MAC (Du et al, 2007), DW-MAC (Sun et al, 2008), AS-MAC (Zhao et al, 2010), and SPEED--MAC (Choi et al, 2010)	End-to-end delay decrease	EIncrease when collision and transmission errors during reservation window	Routing	Reserve the channel along the path toward the sink in advance	Bidirectional, random topology, unicast, broadcast

1.7.2.3 *POWER-SAVING MODES OF OPERATION*

- Sensor nodes communicate using short data packets.
- The shorter the packets, the more dominance of startup energy.
- Operation in a power-saving mode is energy efficient only if the time spent in that mode is greater than a certain threshold.

1.7.2.4 *ERROR CONTROL*

- Error control modes in communication networks (additional retransmission energy cost), forward error correction, automatic repeat request
- Simple error control codes with low-complexity encoding and decoding might present the best solutions for sensor networks.

1.7.3   NETWORK LAYER

- Power efficiency is always an important consideration.
- Sensor networks are mostly data-centric.
- Data aggregation is useful only when it does not hinder the collaborative effort of the sensor nodes.
- An ideal sensor network has attribute-based addressing and location awareness.

1.7.4   TRANSPORT LAYER

- This layer is especially needed when the system is planned to be accessed through Internet or other external networks.
- Transmission control protocol/user datagram protocol-type protocols meet most requirements (not based on global addressing).
- Little attempt thus far to propose a scheme or to discuss the issues related to the transport layer of a sensor network in literature.

1.7.5   APPLICATION LAYER

Management protocol makes the hardware and software of the lower layers transparent to the sensor network management applications.

- Sensor management protocol
- Task assignment and data advertisement protocol
- Sensor query and data dissemination protocol

1.8   CONCLUSIONS

A sensor network is composed of several sensor nodes responsible for collecting information about the environment in which they are deployed and to transmit them to a particular site. It is characterized by high flexibility and fault tolerance, a reduced price, and with fast network deployment. Owing to these characteristics, the scope of sensor networks has expanded to include most areas such as industry, research, the environment, or medicine. It seems clear that they will have new effects in our lives every day and may change significantly our worldview.

However, the realization of sensor networks must satisfy some constraints among which are: energy consumption, the topology change, high-density networks, and so forth. These constraints are double-edged. Indeed, they can satisfy larger projects. However, they make the implementation of a sensor network application challenging.

Clustering is a well-known technique for grouping nodes that are close to each other in the network and to reduce the useful energy consumption. Therefore, we propose a distributed and safe weighted clustering algorithm for mobile WSNs using a combination of five metrics. Among these metrics, the behavioral-level metric lies which promote a safe choice of a cluster head in the sense where this last one will never be a malicious node. In next chapter, we will study several clustering techniques designed for ad hoc and sensor networks. For a successful clustering approach, the need for a powerful mechanism to safely elect a cluster head remains a challenging task in many research works that take into account the mobility of the network. Therefore, we try to propose a new approach based on clustering algorithm to deal with this problem.

KEYWORDS

- **wireless sensor networks**
- **base station**
- **medium access control**
- **power-saving mode**
- **data processing center**

REFERENCES

1. Akyildiz, I. F.; Su, W.; Sankarasubramaniam, Y.; Cayirci, E. Wireless Sensor Networks: A Survey. *Comput. Networks* **2002**, 38(4), 393–422.
2. Bouabdellah, K. Problématique de la consommation d'énergie dans les réseaux de Capteurs sans fil, Université d'Oran. Thèse de doctorat, 2009.
3. Carman, D.; Kruus, P.; Matt, B. J. Constraints and Approaches for Distributed Sensor Network Security. Technical Report 0–010, NAI Labs, 2000.

4. Castellucia, C. La Sécurité des Capteurs et Réseaux de Capteurs, INRIA, juin 2008.
5. Chalel, Y.; Bettahar, H.; Bouabdellah, A. Les Réseaux de capteurs (WSNs: Wireless Sensor Networks); Université de Technologie de Compiègne: France, 2008.
6. Conti, M. Secure Wireless Sensor Networks Threats and Solutions. *Advances in Information Security*; Springer, 2016, Vol 65; ISSN 1568–2633.
7. Crossbow Technology Inc. http://www.xbow.com, 2008.
8. Doudou, M.; Djenouri, D.; Badache, N.; Bouabdallah, A. Synchronous Contention-Based MAC Protocols for Delay-Sensitive Wireless Sensor Networks: A Review and Taxonomy. *J. Network Comput. Appl.*, (*Elsevier*) **2014**, *38*, 172–184.
9. Hill, J. L. *System Architecture for Wireless Sensor Networks*; University of California: Berkeley, spring, 2003.
10. Jurdak, R. *Wireless Ad Hoc and Sensor Networks: A Cross-Layer Design Perspective*; University College Dublin, 2007.
11. Mohamed, L. Diffusion et couverture basées sur le clustering dans les réseaux de capteurs: application à la domotique. Thèse de doctorat, 2009.
12. Walters, J. P.; Liang, Z.; Shi, W.; Chaudhary, V. *Wireless Sensor Network Security: A Survey*; Department of Computer Science Wayne State University, 2006.

CHAPTER 2

CLUSTERING TECHNIQUES

2.1 INTRODUCTION

The clustering in sensor networks is an effective way to minimize the energy consumption in a cluster by running aggregate functions and data fusion in order to reduce the number of messages transmitted to the base station. Recently, several clustering techniques have been proposed to address the challenges defined in sensor networks.

These techniques aim to maintain the information of the network topology, reducing the overhead generated by the discovery of routes and minimize energy consumption taking into account the specificity of these networks.

They are in most cases oriented energy efficient. They aim to extend the lifetime of the network, and in some cases they are oriented Quality of service. In this chapter, we present the aim techniques clustering proposed for sensor networks in the literature.

2.2 DEFINITION

Clustering mean grouping nodes which are close to each other according to a specific metric or a combination of metrics, and forms a virtual topology. Clustering has been widely studied in ad hoc networks[1-9]. Recently, it has been used in wireless sensor networks (WSNs)[10-13] where the purpose in general is to reduce useful energy consumption and routing overhead.

Clusters are generally identified by a particular node called cluster head (CH). This enables coordination between members of the clusters; it aggregates their collected data and transmits them to the base station. It was selected to play that role in a very particular metric or combination of metrics.

Before presenting the different algorithms based on clustering, it is necessary to present the concepts related to this approach.

A wireless network might be modeled by a graph $G=(V, E)$, where V stands for nodes set, that is, mobile devices and E represents existing connections between these nodes.

The process of clustering is to a virtual cutting of V into a plurality of groups close geographically $\{V_1, V_2, V_3, ..., V_k\}$ such that:

$$V = \bigcup_{i=1}^{k} V_i$$

where each subset V_i induces a connected subgraph of G or a connected component of the graph G. The set of these components may form a reduced graph (or components graph) $G'=V' E'$ of G where:

- Nodes $V_i \in V$ correspond to the connected components of G;
- E' contains the edge (V'_i, V'_j) and only if it exists in the graph G an edge of a summit $u_i \in V_i$ to a summit $u_j \in V_j$.

These groups are called "clusters" and they are not necessarily disjoint. Each cluster is identified by a particular node called "CH." The choice of the CH is based on a specific metric or a combination of metrics such as the identifier, degree, energy, K-density, mobility, and so forth.

The effectiveness of a clustering is evaluated in terms of the number of clusters formed and the stability of the clusters based on the mobility of the nodes. The process of clustering is primarily intended to optimize maintenance information of the network topology and deduct overhead diffusion for the discovery of paths.

2.3   CLUSTERS FORMATION

There are several methods of clusters formation. The most widespread[14] one is as follows:

1. Each node must know its neighbors through "hello messages."
2. Each node makes the decision according to his local knowledge of the topology to be a CH or not.

3. The node selected as CH broadcasts its status neighborhood and invites his neighbors who are not yet affiliated with other clusters to join him.

2.3.1   CLUSTER HEAD ELECTION

The phase of CH election also called setup phase uses a specific metric or combination of metrics for each node as the highest-degree/lowest degree in its neighboring, the degree of connectivity, power transmission, the remaining energy, or mobility or a weight that represents a combination of several metrics.

2.3.2   INTRACLUSTER COMMUNICATION AND INTERCLUSTER

Each CH supports communications within the cluster and maintains the routing information allowing it to join the CH. Moreover, as the CHs are not directly connected, the gateway nodes are also elected and used for communication between CHs.

2.3.3   CLUSTERS MAINTENANCE

In order to adapt to changes in network topology, an updated cluster is dynamically performed if a CH or a member of a cluster C_i migrates to another C_j. Furthermore, the CH retains its status as long as possible, even if it does not have the maximum weight in its own cluster. It will lose its role once its battery is turned off or exhausted.

2.4   SOME APPROACHES TO CLUSTERING

In this section, we present and analyze the main building of cluster algorithms in sensor networks and ad hoc networks. These algorithms of clustering gather the clusters formation method, the criteria used to select the CH and maintenance of clusters. The detailed analysis of these algorithms allows us to analyze its limitations and profits from their advantages and enhance its limitations. Then, we review some approaches of clustering in ad hoc networks and WSNs. We summarize all the examined algorithms in Table 2.1.

Using this table, we want to further clarify the main features of different types of clustering approaches. Furthermore, this way of presenting makes easier the comparison between various techniques. The first column specifies the algorithm name. The second column specifies the authors of the proposed approach. The other columns are devoted to the most representative grouping functions with which we can easily see the basic differences and common aspects of different approaches.

The meaning of each column is explained in the following:

- Lifespan: This field specifies the metric used by the authors to evaluate the lifetime of the network. In our case, we considered that the network will be invalid when the nodes in the neighborhood of the sink exhaust their energy.
- Energy efficiency: This field indicates if the proposed algorithm guaranty energy saving in order to extend the lifespan of the network or not.[16]
- Energy consumption in a sensor node can be due to either "useful" or "wasteful" sources. Useful energy consumption can be due to transmitting/receiving data, processing query requests, and forwarding queries/data to neighboring nodes. Wasteful energy consumption can be due to Idle listening to the media, retransmitting due to packet collisions, overhearing, and generating/handling control packets.
- Load balancing: This field specifies the parameters enabling to generate a reduced number of stable and balanced clusters in each scheme.
- Security: This field specifies the robustness of the scheme to detect specific misbehavior in the network and, the very important thing, the type of these attacks.[15]

In what follows, we present the list of basic items that were used to list the characteristics of different algorithms in Table 2.1

BASE PAPERS

1. Chatterjee, M.; Das, S. K.; Turgut, D., "WCA: A Weighted Clustering Algorithm for Mobile ad Hoc Networks," Cluster Comput. 2002, 5(2), 193–204.

2. Choi, W.; Woo, M., A Distributed Weighted Clustering Algorithm for Mobile ad Hoc Networks, in Proceedings of Advanced International Conference on Telecommunications and International Conference on Internet and Web Applications and Services (AICT-ICIW'06), Guadeloupe, French Caribbean, 2006; pp 73–73.

3. Zabian, A.; Ibrahim, A.; Al-Kalani, F., Dynamic Head Cluster Election Algorithm for Clustered Ad-Hoc Networks. J. Comput. Sci. 2008, 4(1), 42–50.

15. da Silva, A. P. R.; Martins, M. H.; Rocha, B. P.; Loureiro, A. A.; Ruiz, L. B.; Wong, H. C. Decentralized Intrusion Detection in Wireless Sensor Networks, In Proceedings of the 1st ACM International Workshop on Quality of Service & Security in Wireless and Mobile Networks, Montreal, Canada, 2005; pp 16–23.

16. Lehsaini, M.; Guyennet, H.; Feham, M. An Efficient Cluster-Based Self-Organisation Algorithm for Wireless Sensor Networks. Int. J. Sens. Networks, 2010, 7(1), 85–94.

The metrics used in our algorithm; Mobility (M_i), Connectivity (), Residual Energy (), Behavior Level of Node (BL_i), Distance between node and its neighbors (D_i) are explained in Chapter 4.

2.4.1   LOWEST-ID HEURISTIC

The Lowest-ID, also known as identifier-based clustering, was originally proposed by Baker and Ephremides.[17] This heuristic assigns a unique "ID" to each node and chooses the node with the minimum "ID" as a CH. Thus, the "IDs" of the neighbors of the CH will be higher than that of the CH. However, the CH can delegate its responsibility to the next node with the minimum "ID" in its cluster. A node is called a gateway if it lies within the transmission range of two or more CHs. Gateway nodes are generally used for routing between clusters.

TABLE 2.1 A Survey of Clustering Schemes.

Clustering Schemes	Authors	Weighted	M_i	D_i	Er_i	C_i	BL_i	Lifespan	Energy Efficiency	Load balancing	Security
LEACH	(Heinzelman et al., 2000)	✗	✗	✗	✔	✗	✗	Poor	Very poor	Medium	Poor
Lowest-ID	(Baker and Ephremides, 1981)	✗	✗	✗	✗	✗	✗	Poor	Very poor	Very poor	Poor
Highest-ID	(Gerla and Tsai, 1995)	✗	✗	✗	✗	✗	✗	Poor	Very poor	Very poor	Poor
Single metric (Energy)	(Ye and Chong., 2005)	✗	✗	✗	✔	✗	✗	Poor	Poor	Poor	Poor
Single metric (Stability)	(Er and Seah., 2004)	✗	✔	✗	✗	✗	✗	Poor	Poor	Poor	Poor
Single metric (Connectivity)	(Ye and Chong., 2005)	✗	✗	✗	✗	✔	✗	Poor	Poor	Medium	Poor
WCA	(Chatterjee et al., 2002)	✔	✔	✔	✔	✔	✗	Poor	Poor	Poor	Poor
DWCA	(Choi and Woo., 2002)	✔	✔	✔	✔	✔	✗	Medium	Medium	Medium	Poor
DHCEA	(Zabian et al., 2008)	✔	✔	✗	✔	✔	✗	Medium	Medium	Medium	Poor
DEECA	(Safa et al., 2008)	✔	✔	✗	✔	✗	✗	Good	High	Poor	Poor
CBTRP	(Safa et al., 2010)	✔	✔	✗	✔	✔	✔	Good	Good	Poor	High
ECSA	(Lehsaini et al., 2010)	✔	✗	✗	✔	✔	✗	Good	High	High	Poor
RECA	(Elhdhili et al., 2008)	✔	✔	✔	✔	✔	✔	Medium	Medium	Medium	Good
SCAR	(Yu and Zang., 2012)	✔	✔	✗	✔	✔	✔	Poor	Poor	Poor	High
SDCA	(Benahmed et al., 2013)	✔	✔	✔	✔	✔	✔	Medium	Poor	Medium	Good
HEED	(Younis et al., 2004)	✗	✗	✗	✔	✔	✗	Good	Good	Medium	Poor
fuzzy based k-Hop	(Jain and Reddy., 2015)	✗	✗	✗	✔	✗	✗	High	High	Medium	Poor
EECS	(Ye et al., 2005)	✗	✗	✗	✔	✔	✗	Good	Good	Medium	Poor
ES-WCA	(Dahane et al., 2015)	✔	✔	✔	✔	✔	✔	Good	High	High	High
EWCA	(Li et al., 2009)	✔	✔	✔	✔	✔	✗	Medium	Medium	Medium	Poor
FWCABP	(Hussein et al., 2008)	✔	✔	✔	✔	✔	✗	Medium	Medium	Medium	Poor

The lowest-ID algorithm is not convenient for clustering in WSNs because the node with the smallest identifier in its neighborhood can become CH even if its energy supply is low (cf. Fig. 2.1).

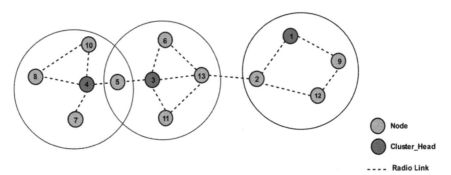

FIGURE 2.1 Cluster formation where each node announce cluster head (CH) or not based on its "ID" and those of its neighbors.

2.4.2 HIGHEST-DEGREE HEURISTIC

The highest degree, also known as connectivity-based clustering, was originally proposed by Gerla and Tsai[18] in which the degree of a node is computed based on its distance from others. Each node broadcasts its "ID" to the nodes that are within its transmission range.

A node x is considered to be a neighbor of another node y if x lies within the transmission range of y. The node with a maximum number of neighbors (i.e., maximum degree) is chosen as a CH and any tie is broken by the unique node IDs.

The neighbors of a CH become members of that cluster and can no longer participate in the election process. Since no CHs are directly linked, only one CH is allowed per cluster. Any two nodes in a cluster are at most two-hops away since the CH is directly linked to each of its neighbors in the cluster.

Basically, each node either becomes a CH or remains an ordinary node (neighbor of a CH).

Experiments demonstrate that the system has a low rate of CH change but the throughput is low under the highest-degree heuristic. The re-affiliation count of nodes is high due to node movements and as a result, the highest-degree node (the current CH) may not be reelected to be a CH even if it loses one neighbor.

2.4.3 CLUSTER FORMATION BASED ON ENERGY

Mobile nodes in a Mobile ad hoc Network normally depend on battery power supply during operation; hence, the energy limitation possesses a real challenge for network performance. There should be some mechanism to reduce the energy consumption of the node in order to prolong the network lifespan. Moreover, a CH bears extra work compared with ordinary members, and it is more likely to die early because of excessive energy consumption that may cause network partition and hence the communication interruption. For this reason, it is important to balance the energy consumption among mobile nodes to avoid node failure.[14]

2.4.3.1   CLUSTER FORMATION

- The depletion of energy of CH causes the breakdown of the cluster.
- The node with the highest residual energy is chosen as CH among the neighboring nodes so as to prolong the cluster lifetime. In case of ties, the lowest ID wins over.

2.4.3.2   CLUSTER MERGING

- Cluster energy = Average residual energy of nodes in that cluster.
- The gateway node will choose the cluster that results in the highest improvement percentage of the cluster energy or the lowest degrading percentage of the cluster energy when it is included in that cluster.
- Cluster energy with the addition of a gateway node = (cluster energy + residual node energy)/2.
- However, merging the percentage improvement of energy should be above the threshold value (10%).
- If the percentage decrease in cluster energy falls below 50%, cluster should not be merged.

2.4.3.3   CLUSTER SPLITTING

- If the rate of energy loss of primary CH (PCH) exceeds the threshold then the cluster gets split.
- Initially, every node exchanges its node ID and residual energy value.
- Let us suppose node 1 initiates the "hello" message and its neighbors also initiate the "hello" message to their neighbors and this "hello" message propagate throughout the network (see Fig. 2.2).
- Comparing its neighbors, node 3 having highest residual energy value broadcast its decision to be the CH.

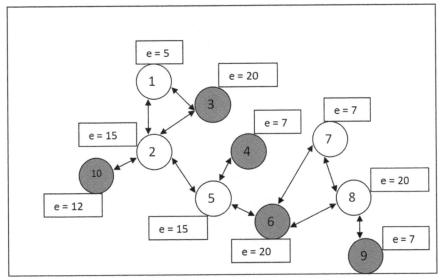

FIGURE 2.2 Example network where "e" represents the residual energy for the node.
Source: Reprinted from Sanjaya Gajurel, "Multi-Criteria Clustering," Case Western Reserve University Cleveland, OH, USA, Technical Report, 2006. Published under the Creative Commons License.

- Nodes 1 and 2 which are the neighbors of node 3 join as the members of cluster with CH node 3.
- Node 2 is already a member of a cluster and it has CH ID information so it cannot be the CH in the next stage. At the second stage, node 6 is eligible to be a CH having the highest value of residual energy.
- Though node 6 and node 8 has the same residual energy, the tie is broken by lowest ID.
- Node 6 broadcasts it's CH information and nodes 5, 7, and 8 become its members.
- Nodes 4, 9, and 10 not hearing the CH broadcasts for a certain time period declare themselves as CH. CH 4 could have a cluster member 5, CH 10 could have a cluster member 2, and CH 9 could have a cluster member 8. However, those nodes are already the members of other clusters and do not get registered to the new cluster. Those nodes are the gateway nodes.
- The flag is set to 1 for the gateway nodes.

- Node 1 and node 8 are selected as the secondary CH (SCH) for the cluster 1 and cluster 3, respectively. The completed cluster formation process results in the cluster structure shown in Figure 2.3.

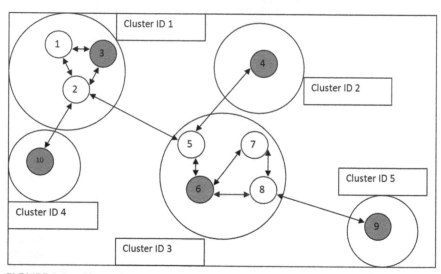

FIGURE 2.3   Cluster formation.

Source: Reprinted from Sanjaya Gajurel, "Multi-Criteria Clustering," Case Western Reserve University Cleveland, OH, USA, Technical Report, 2006. Published under the Creative Commons License.

2.4.3.4   MERGING OF CLUSTERS

Gateway node of cluster 1, that is, node 2 can either merge to cluster 4 or 3 or stay in the same cluster depending on the increment or decrement percentage of average energy of the cluster with the entry or exit of the gateway node 2.

- Without including gateway node 2
  Cluster energy of cluster 1 is: $25/2 = 12.5$
  Cluster energy of cluster 4 is: 12
  Cluster energy of cluster 3 is: $62/4 = 15.5$
- Including gateway node 2
  Cluster energy of cluster 1 is $40/3 = 13.3$
  Cluster energy of cluster 4 is $27/2 = 13.5$

Cluster energy of cluster 3 is $77/5 = 15.4$
- Increment/decrement percentage
  Cluster 1: $(13.3–12.5)/12.5 \times 100 = 6.4\%$ (increment)
  Cluster 4: $(13.5–12)/12 \times 100 = 12.5\%$ (increment)
  Cluster 3: $(15.4–15.5)/15.5 \times 100 = -0.6\%$ (decrement)

Increment percentage of energy for cluster 4 is the highest and above the threshold, so cluster 1 gets merged to cluster 4. Node 3 having highest residual energy value among two acts as PCH and node 10 acts as an SCH. The cluster structure is shown in Figure 2.4. Cluster 3 with gateway node 5 has 3 choices to merge—stays as a separate cluster, merge with cluster 1 or 2.

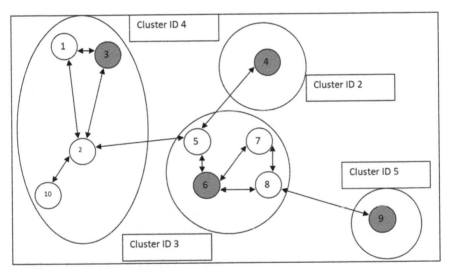

FIGURE 2.4   First phase of cluster merging.
Source: Reprinted from Sanjaya Gajurel, "Multi-Criteria Clustering," Case Western Reserve University Cleveland, OH, USA, Technical Report, 2006. Published under the Creative Commons License.

- Without including gateway node 5
  Cluster energy of cluster 4 is: $52/4 = 13$
  Cluster energy of cluster 3 is: $47/3 = 15.6$
  Cluster energy of cluster 2 is: $7 = 7$

- Including gateway node 5
  Cluster energy of cluster 4 is: $67/5 = 15.4$

Cluster energy of cluster 3 is: 62/4 = 15.5
Cluster energy of cluster 2 is: 22/2 = 11

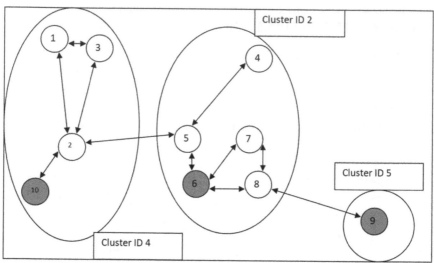

FIGURE 2.5   Second phase of merging.
Source: Reprinted from Sanjaya Gajurel, "Multi-Criteria Clustering," Case Western
Reserve University Cleveland, OH, USA, Technical Report, 2006. Published under the
Creative Commons License.

Cluster 3 merges with cluster 2, increased percent (= 36%) above the
threshold value; the previous cluster having highest energy takes over as
PCH and the other as SCH. The cluster structure takes the form as shown
in Figure 2.4.

In the final phase, cluster 2 can either stay intact or merge with cluster
5 (Fig 2.5).

• Without including gateway node 8
  Average energy of cluster 2 is: 49/4 = 12.25
  Average energy of cluster 5 is: 47/3 = 7

• Including gateway node 5
  Average energy of cluster 2 is: 69/5 = 13.8
  Average energy of cluster 3 is: 27/2 = 13.5

Cluster 2 gets merged with cluster 5, the increased percent in average
energy being 93%; CH 6 having the highest residual energy becomes a

PCH and the other CH 9 becomes SCH. The final cluster structure after merging is shown in Figure 2.6.

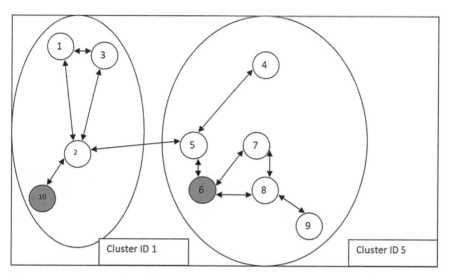

FIGURE 2.6   Cluster structure after the merging process.

Source: Reprinted from Sanjaya Gajurel, "Multi-Criteria Clustering," Case Western Reserve University Cleveland, OH, USA, Technical Report, 2006. Published under the Creative Commons License.

2.4.4   CLUSTER FORMATION BASED ON CONNECTIVITY

The idea for selecting most connected CH is to make the CH connected to its members even when some of the links are broken. The connectivity of the node is defined in terms of the "degree of the node," that is, the number of nodes that node is connected with. The node having the highest degree among its neighbors is chosen as the CH.

Pure connectivity-based algorithm does not work properly as there can be many neighbors having the same degree. The ties among the neighbors are arbitrated by using the lowest ID; node having the lowest ID wins over. The SCH is also elected for each cluster during the election of PCH. The cluster so formed will be 1-hop cluster. However, the merging of clusters occurs under certain conditions that result in variable size cluster formation.[14]

2.4.4.1   MERGING OF CLUSTERS

- No cluster can merge with the single node cluster; however, a single node cluster should merge with one of the clusters with which it is in communication range.
- Cluster connectivity = number of links in the cluster/maximum number of possible links:

$$= \frac{(1+x)}{n.(n+1)}$$

where n is the number of nodes in the cluster.

- The gateway node will choose the cluster that results in the largest improvement percentage of the cluster connectivity or the lowest degrading percentage of the cluster connectivity when it is included in that cluster.
- Cluster connectivity with the addition of a gateway node = number of links in the cluster with gateway node/maximum number of possible links with gateway node:

$$= \frac{(1+x)}{n.(n+1)}$$

Where x is the number of addition links created with the addition of that node.

- However, merging the percentage improvement of connectivity should be above some threshold value (Fig. 2.7).
- If the percentage decrease in cluster connectivity falls below 50% with the addition of the node, cluster should not be merged.

2.4.4.2   SPLITTING OF CLUSTERS

- Cluster connectivity $\leq$ threshold connectivity
- Threshold connectivity (on the basis of Fig. 2.1)=4/(5/2 * (5−1))=0.4

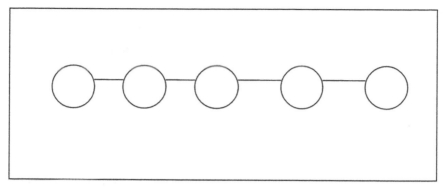

FIGURE 2.7   Scenario for calculating the threshold value of connectivity.

Source: Reprinted from Sanjaya Gajurel, "Multi-Criteria Clustering," Case Western Reserve University Cleveland, OH, USA, Technical Report, 2006. Published under the Creative Commons License.

Example: Cluster formation (This algorithm different from others—not simultaneous cluster formation but sequential).

Let us consider the example network shown in Figure 2.8.

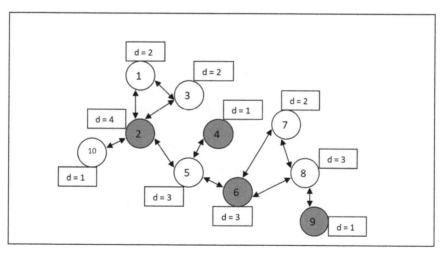

FIGURE 2.8   Example network where "d" is the degree of a node.

Source: Reprinted from Sanjaya Gajurel, "Multi-Criteria Clustering," Case Western Reserve University Cleveland, OH, USA, Technical Report, 2006. Published under the Creative Commons License.

- Our cluster formation is a one-hop cluster formation but merging of clusters results in variable sized clusters structure.
- Initially, every node exchanges its node ID and neighbor list from which degree "d" is determined.
- In Figure 2.9, let's suppose node 1 initiates the "hello" message. Now, its neighbors also initiate the "hello" message to their neighbors and this "hello" message propagates throughout the network.
- Comparing its neighbors, node 2 having the highest degree broadcasts its decision to be the CH.
- Nodes 1, 3, 5, and 10 which are the neighbors of node 2 join as the members of cluster with CH node 2.
- Node 5 is already a member of a cluster and it has CH ID information so it cannot be the CH in the next stage. Therefore, at the second stage, node 6 is eligible to be a CH.
- Though node 6 and node 8 have the same degree, the tie is broken by lowest ID.
- Node 6 broadcasts its CH information and nodes 7 and 8 become its members.
- Node 4 and node 9 not hearing the CH broadcasts for a certain time period, declare themselves as CH.
- Node 5 could have been the cluster member for clusters 1, 2, and 3 and node 8 could have been the cluster member for Clusters 3 and 4. These nodes are the gateway nodes. However, the gateway node is registered as a member to only one cluster; the initially formed cluster.
- The flag is set to 1 for the gateway nodes.
- Node 1 and node 8 are selected as the SCH for the cluster 1 and cluster 3 respectively.
- The final formation of cluster is shown in Figure 2.11.

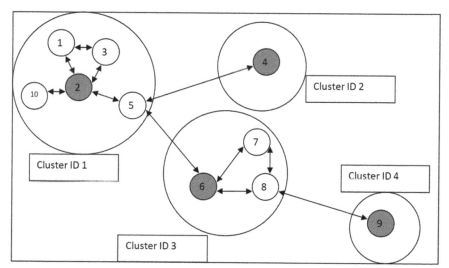

FIGURE 2.9 Cluster formation based on connectivity.

Source: Reprinted from Sanjaya Gajurel, "Multi-Criteria Clustering," Case Western Reserve University Cleveland, OH, USA, Technical Report, 2006. Published under the Creative Commons License.

Example: Merging of clusters
The gateway node 5 initiates the merging with either cluster 2 or cluster 3.

- Cluster connectivity without gateway node 5
  For cluster 1, Cluster connectivity = 0.67
  For cluster 3, Cluster connectivity = 1.00

- Cluster connectivity with gateway node 5
  For cluster 1, Cluster connectivity = 0.50
  For cluster 3, Cluster connectivity = 0.67

- Percentage difference
  Cluster 1: −25.4%
  Cluster 3: −33.33%

Cluster 1 is not merging with cluster 3 as the decrement percent is less in cluster 1 than cluster 3.

Cluster 2 being a single node cluster simply merges with cluster 1. The cluster structure so formed is shown in Figure 2.10.

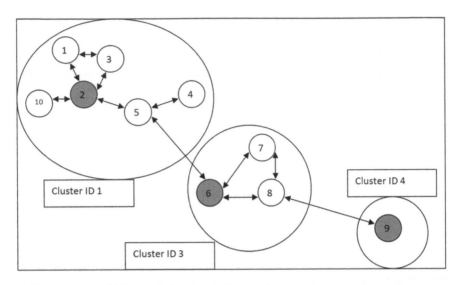

FIGURE 2.10   Initial step of cluster merging process.

Source: Reprinted from Sanjaya Gajurel, "Multi-Criteria Clustering," Case Western Reserve University Cleveland, OH, USA, Technical Report, 2006. Published under the Creative Commons License.

- Cluster connectivity without gateway node 6
  For cluster 1, Cluster connectivity = 0.4
  For cluster 3, Cluster connectivity = 1.00

- Cluster connectivity with gateway node 6
  For cluster 1, Cluster connectivity = 0.33
  For cluster 3, Cluster connectivity = 1.00

- Percentage Difference
  Cluster 1: −16.67%
  Cluster 3: 0.00%

  Cluster 3 is not merging with cluster 1.
  Cluster 4 being a single node cluster simply merges with cluster 3.
  The final cluster structure is shown in Figure 2.11.

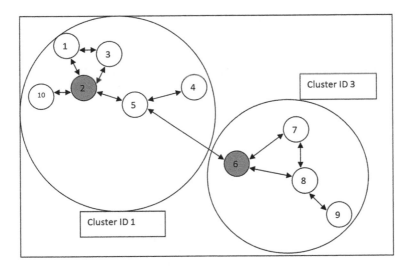

FIGURE 2.11 Final cluster structure.

Source: Reprinted from Sanjaya Gajurel, "Multi-Criteria Clustering," Case Western Reserve University Cleveland, OH, USA, Technical Report, 2006. Published under the Creative Commons License.

Example: Splitting of clusters

Let's suppose a new node 11 joins the cluster 1 as shown in Figure 2.12.

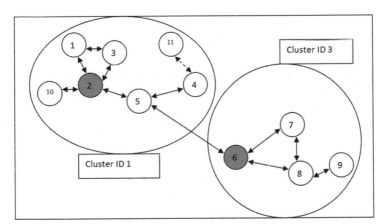

FIGURE 2.12 Scenario when a new node joins the cluster and its effect on cluster connectivity.

Source: Reprinted from Sanjaya Gajurel, "Multi-Criteria Clustering," Case Western Reserve University Cleveland, OH, USA, Technical Report, 2006. Published under the Creative Commons License.

Cluster connectivity of cluster 1 = 0.38

Cluster connectivity of cluster 3 = 0.67

Since the cluster connectivity of cluster 1 is below the threshold value of 0.4, it should be split as shown in Figure 2.13. The PCH 2 becomes the CH for cluster 1 and SCH 4 becomes the CH for new cluster 4.

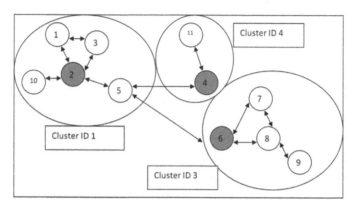

FIGURE 2.13   Splitting of cluster.

Source: Reprinted from Sanjaya Gajurel, "Multi-Criteria Clustering," Case Western Reserve University Cleveland, OH, USA, Technical Report, 2006. Published under the Creative Commons License.

2.4.5 A WEIGHTED CLUSTERING ALGORITHM (WCA) FOR MOBILE AD HOC NETWORKS

Chaterjee et al.[1] proposed an on-demand distributed clustering algorithm for multi-hop packet radio networks. These types of networks, also known as ad hoc networks, are dynamic in nature due to the mobility of nodes. The association and dissociation of nodes to and from clusters perturb the stability of the network topology, and hence a reconfiguration of the system is often unavoidable. However, it is vital to keep the topology stable as long as possible. The CHs form a dominant set in the network, determine the topology and enhance its stability.

The proposed weight-based distributed clustering algorithm takes into consideration the ideal degree, transmission power, mobility, and battery power of mobile nodes. The time required to identify the CHs depends on the diameter of the underlying graph. Authors try to keep the number of nodes in a cluster around a predefined threshold to facilitate the optimal operation of the medium access control protocol. The nonperiodic procedure for CH election is invoked on-demand, and is aimed to reduce the computation and communication costs.

The CHs, operating in "dual" power mode, connects the clusters which help in routing messages from a node to any other node. They observed a trade-off between the uniformity of the load handled by the CHs and the connectivity of the network. Simulation experiments are conducted to evaluate the performance of our algorithm in terms of the number of CHs, re-affiliation frequency, and dominant set updates. Results show that our algorithm performs better than existing ones and is also tunable to different kinds of network conditions (Table 2.2).

TABLE 2.2 Execution of Weighted Clustering Algorithm (WCA)[1]

Node id	d_v step 1	Δ_v step 2	D_v step 3	M_v step 4	P_v step 5	W_v step 6
1	2	0	6	2	1	1.35
2	1	1	4	2	2	1.70
3	1	1	3	3	1	1.50
4	1	1	3	4	2	1.60
5	3	1	9	1	4	2.75
6	1	1	3	2	2	1.50
7	2	0	6	0	0	1.20
8	2	0	7	3	3	1.70
9	4	2	13	2	6	4.40
10	3	1	12	2	7	3.55
11	1	1	3	0	1	1.35
12	2	0	5	3	4	1.35
13	2	0	7	3	2	1.65
14	2	0	5	2	0	1.10
15	1	1	3	4	3	1.65

2.4.6 A DISTRIBUTED WEIGHTED CLUSTERING ALGORITHM (DWCA) FOR MOBILE AD HOC NETWORKS

Author's aims for the clustering process are set to maintain stable clustering structure, to minimize the overhead for the clustering set-up and maintenance and to maximize the lifespan of mobile nodes in the system.

To achieve these goals, they propose a DWCA. The proposed algorithm is an extended version of weighted clustering algorithm (WCA)[1] that utilizes factors like the number of neighbors, mobility, and battery power for election of the CHs. Our algorithm differs from WCA in which it localizes configuration and reconfiguration of clusters and poses restriction on the power requirement of the CHs. The result of simulations shows that the proposed algorithm provides better performance than WCA in terms of the number of re-affiliations, end-to-end throughput, and overheads during the initial clustering setup phase, and the minimum lifespan of nodes.[2]

2.4.6.1   PHASE 1: CLUSTERING SETUP

Initially, each node broadcasts a "hello" message to notify its presence to the neighbors. A "hello" message contains its ID and position value. Each node builds its neighbor list based on the "hello" messages received.

Election of the CHs is based on the weight values of the neighbor nodes. Each node calculates its weight value based on the following factors:

- Degree difference: the difference between the actual number of neighbors and the number of nodes that a CH can handle ideally.
- Sum of distances with all its neighbors using the position value obtained from the "hello" messages.
- Running average of speed.
- Consumed battery power.

These factors are mainly based on the weight factors used in WCA,[1] except the consumed battery power.

2.4.6.2   PHASE 2: CLUSTERS MAINTENANCE

The second phase is the clustering maintenance. There are two situations that invoke the clustering maintenance. One is node movement to the outside of its cluster boundary and the other is excessive battery consumption at a CH. When an ordinary node moves to the outside of its cluster boundary, it is required to find a new CH to affiliate with.

If it finds a new CH, it hands over to the new one. If not, it declares itself as a CH. If node v moves out of the transmission range of its current CH x, it broadcasts a "Find_CH" message. If any CH receives this message, it sends a "CH_Ack" message to node v.

Then node v chooses a CH with the lowest weight value based on the received "CH_Ack" messages and hands over to the chosen CH by sending a "Join_CH" message. If node v does not receive any "CH_Ack" message within a given time period, it declares itself as a CH by sending a cluster message.

a) Move
b) Broadcast a Find_CH message
c) Do not receive any CH_ACK message
d) Become a CH and broadcast a Cluster message

The performance of the proposed ES-WCA algorithm (DWCA) is measured by calculating:

- The number of clusters
- The number of re-affiliations
- End-to-end throughput
- Overhead of packets generated during the initial
- Clustering setup phase
- Minimum lifespan of a node which is the operational time of a node that fails first due to battery exhaustion

SIMULATION RESULTS

The "DWCA" algorithm produces fewer clusters than (WCA).

Figure 2.12 depicts the average number of clusters formed with respect to the total number of nodes in the ad hoc network with different node speeds. The number of clusters increased as the number of nodes and node speed increased. As we can see in Figure 2.12, the proposed algorithm produced less clusters than WCA. When nodes were sparsely located in the network and moved slowly, DWCA produced about 9.8% less clusters than WCA.

The performance difference became smaller if sparsely located nodes moved fast. If the node density increased, DWCA produced constantly less clusters than WCA regardless of node speed. When there were 70 nodes in the ad hoc network, the proposed algorithm produced about 7.5–9.2% less clusters than WCA.

As a result, our algorithm gave better performance in terms of the number of clusters when the node density and node mobility in the network are high.

The result of the average number of re-affiliations due to node mobility is depicted in Figure 2.13. As shown in the figure, when the node speed was not faster than 10 m/s, the number of re-affiliations increased linearly if there were 30 or less nodes in the network for both our algorithm and WCA. As the number of nodes and the mobility of nodes increased, the increasing rate of re-affiliations slowed down. According to the result, DWCA constantly gave fewer re-affiliations than WCA especially for the high node density and low node mobility. When there were 10 nodes in the network, the proposed algorithm produced 34.0% less re-affiliations than WCA for the node speed of 3 m/s. When the number of nodes was increased to 70, our algorithm gave 43.2% fewer re-affiliations than WCA for the node speed of 3 m/s.

When nodes moved with the speed of 30 m/s, our algorithm produced 23.8 and 12.5% less re-affiliations than WCA for the number of nodes 10 and 70, respectively. The benefit of decreasing number of re-affiliations mainly comes from the localized cluster maintenance in our algorithm.

2.4.7   DYNAMIC HEAD CLUSTER ELECTION ALGORITHM FOR CLUSTERED AD HOC NETWORKS

Zabian et al.[3] proposed a distributed clustering and leader election mechanism for ad hoc mobile networks, in which the leader is a mobile node. Their results show that, in the case of leader mobility, the time needed to elect a new leader is smaller than the time needed for significant topological change in the network.

2.4.7.1   ALGORITHM DESCRIPTION

Before describing the algorithm it is useful to define some notations used in the algorithm description:

- Mobile terminal (MT): refers to a MT.
- HC: refers to the head cluster.
- P (HC): refers to the power of the HC
- D (HC) refers to the number of neighbors of the HC.
- N: represents the number of nodes in the network.

In constructing our network, the DHCEA uses three types of messages that are:

- Leader election message (LE) that is sent in the network when the nodes start working or when the leader is moved to another area.
- The second type of messages is the Neighbor Discovering message (ND) that is sent by the HC or any discovered node in the cluster formation phase to collect information about the nodes in its environment.
- The last type of messages is the Acknowledgement message (ACK) that will be sent periodically by the HC to maintain information about the network.

The algorithm runs in three phases: the "head cluster election phase," the "cluster formation phase," and finally the "network maintenance phase."

2.4.7.2   PHASE 1: HEAD CLUSTER ELECTION

- A node U chosen randomly starts discovering its environment by broadcasting a LE to which is attached its power level and its identity and sets a timeout t. t is the time needed to a radio waves to travel in an environment without obstacles a distance equal to the maximum area covered by the transmission range ($t = 2 \, 1 + \varepsilon$; where l is the maximum distance covered by the transmission range and ε is the processing time at the destination will be ignored) and $r = 1. \, d^{\alpha}$ [3]. Each node hears the LE and wants to join the cluster; it must send a response message that contains its power level and its identity. If U does not receive any response, it waits a time t and sends another LE message until the reception of a response or until hearing an LE message coming from other nodes.
- If two nodes start the leader election process at the same time and the two nodes can hear each other, one stopped and the other

continues the election process. If the two nodes cannot hear each other, each node selects a leader in its environment and at the end of the process, only one as leader is chosen (step 6, phase1).

- All the response messages received is inserted in a queue, from which U sorts one node with higher power level as leader.
- If there are two nodes in the queue with the same power level (higher level) the node with higher number of neighbors is chosen.
- Then U distributes the leader identity in the cluster.
- If there is another leader in the same cluster (for scalability issue we assume that not all the nodes in the cluster are reachable in only one hop), the two headers will be compared given the higher power level and given the proximity to more nodes in the cluster (a header is supposed to be close to a large part of nodes of the cluster). If the two HCs have the same power level, the head cluster with large set of neighbors is chosen, otherwise, one HC is chosen as HC and the other will be considered as substitute HC (considering that the HC is a mobile node and can change location with the time). The power level P varies from 1−k, where k is the maximum power. The unreachable node by the header can be reached by multi-hops given another MT. Figure 2.14 shows the flowchart that describes the first phase of the algorithm that is head cluster election.

2.4.7.3   PHASE 2: CLUSTER FORMATION PHASE

In this phase:

- The elected HC broadcasts an ND message to construct its cluster. All the nodes that hear the ND message and want to join it, must send a response message. Based on which the HC assigns to them their identity in the network.
- The ND message can be heard by two types of nodes: nodes not yet discovered (new placed in this area, or were down and resumed activity) that means do not have a parent. If it wants to join this HC, it must send a response message. The second type of nodes is nodes that have been discovered and has its parent that means belonging to another cluster. This type of nodes represent important nodes in the routing process because it represents intermediate nodes; such

node has the liberty to take this link as additional link or to ignore the ND message.

- Based on the responses received, the HC constructs its routing table in which is inserted the identity of the nodes connected to it and its power level (from the same cluster or from the other clusters) and will be denoted as children to which is assigned their identities.
- Steps 1 and 3 are repeated in each cluster until the construction of the entire network.
- If an HC node hears the ND message coming from an MT or from another HC and it wants to collaborate, it must send a response message. In the case where the ND messages were exchanged between HCs, each HC adds in its routing table a new row in which all the HCs connected to it are inserted (that means that they are in its transmission range).
- If a node (MT) receives an ND message from two HCs, it can choose only one as HC from which it takes its identity and the other HC will be reserved as backup links. Such nodes represent a high importance in our system because it will be considered as intermediate nodes that can carry out a communication between two clusters with low cost.

For example, in Figure 2.15, both nodes "011" and "043" represent intermediate nodes between two clusters and are consecutively G1 then G2, and G4 then G3.

The address assignment: One of the roles assigned to the HC is to assign a unique identity to the nodes of its cluster. The addresses assigned are composed on three components "XYZ", where each one can be a 4 bits character coding the numbers from 0–9 (for more details, see in ref 3).

2.4.7.4   PHASE 3: NETWORK MAINTENANCE

If some changes happened in the network topology, they will be attached in a piggyback to the ACK, so the HC sends periodically ACKs to collect information about its cluster. Concerning mobility management, we can find the following two cases as explained in flowchart below (Fig. 2.14).

- Election of one HC without substitute so the Head Cluster Election phase must be restarted.
- Election of a substitute HC.

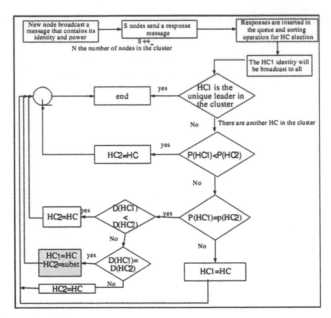

FIGURE 2.14   Head cluster election.
Source: Reprinted from ref [3]. Used with permission of the Creative Commons License.
https://creativecommons.org/licenses/by/4.0/.

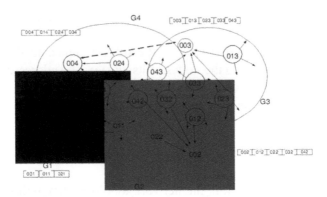

FIGURE 2.15   (See color insert.) Phase 2, cluster formation phase.
Source: Reprinted from ref [3]. Used with permission of the Creative Commons License.
https://creativecommons.org/licenses/by/4.0/.

Figures 2.16A, B, C, and D show a simulation of the first and second phases of DHCEA in a graph of 11 nodes. Figure 2.16A simulates the leader election phase where two nodes randomly start broadcasting an LE message. In part A of Figure 2.16B, the nodes that hear LE message and want to cooperate, send a response message. Part B of the same figure represents the formation of the right part of Figure 2.16C represents how the leaders elected assign the identities to their child and each one constructs its routing table. Then, each discovered node represented by bold point starts discovering its environment. Figure 2.17 shows that τ is independent on the cluster size [3].

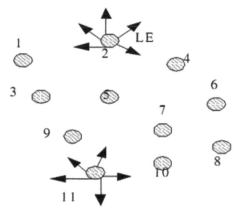

FIGURE 2.16A   Leader election phase.

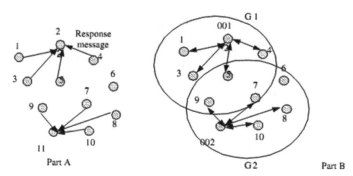

FIGURE 2.16B   Cluster formation.

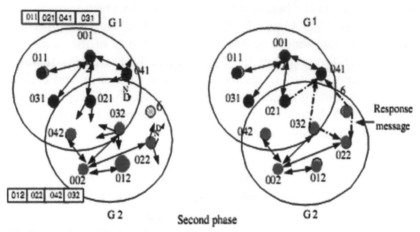

FIGURE 2.16C   Address assignment.

Source: Reprinted from ref [3]. Used with permission of the Creative Commons License.
https://creativecommons.org/licenses/by/4.0/.

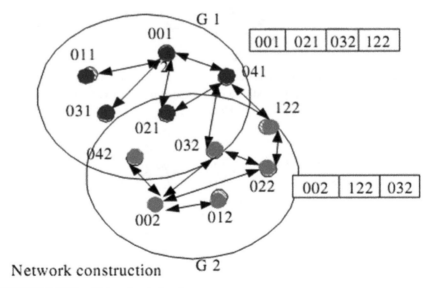

FIGURE 2.16D   Network construction.

Source: Reprinted from ref [3]. Used with permission of the Creative Commons License.
https://creativecommons.org/licenses/by/4.0/.

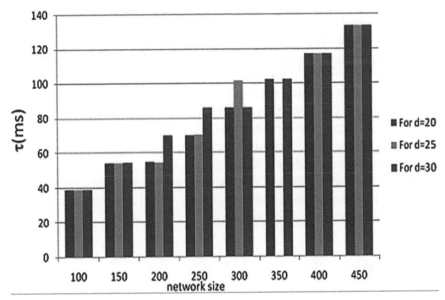

FIGURE 2.17 **(See color insert.)** The variation of the time needed for leader election (τ) with the network size.

Source: Reprinted from ref [3]. Used with permission of the Creative Commons License. https://creativecommons.org/licenses/by/4.0/.

2.5 CONCLUSION

In this chapter, we studied several clustering techniques designed for ad hoc networks and sensor networks. We found that clustering mimics the centralized architecture and drives profits of its advantages in small- to medium-sized networks. It allows spatial reuse of radio frequencies to minimize interference and the additional cost of traffic control. It is well suited for sensor networks since they have low memory to store the entire network topology. We also found that it is necessary to choose one or more metrics for CH election in order to build more or less stable clusters. These metrics should reflect the ability of nodes to act as a CH and network topology. However, the power constraint is considered a strong constraint in sensor networks. Moreover, to ensure load balancing between nodes clusters formed will be homogenous in size. All clusters generated must contain almost the same number of nodes and radius, all cluster members must be most K-hops in their CH in order to reduce traffic control induced by the implementation of routing, service discovery, also the role of the CH will be limited time period and clustering should be distributed.

The main concern of clustering approaches for mobile WSNs is to prolong the battery life of the individual sensors and the network lifetime. For a successful clustering approach, the need for a powerful mechanism to safely elect a CH remains a challenging task in many research works that take into account the mobility of the network. The approach based on the computing of the weight of each node in the network is one of the proposed techniques to deal with this problem. Little effort has been made by other researchers in the last decade to include the security aspect in the clustering mechanism so in the next chapter we will see the security and privacy challenges in wireless networks.

KEYWORDS

- cluster head
- ad hoc networks
- gateway node
- weighted clustering algorithm
- mobile terminal

REFERENCES

1. Chatterjee, M.; Das, S. K.; Turgut, D., "WCA: A Weighted Clustering Algorithm for Mobile ad Hoc Networks," *Cluster Comput.* **2002**, 5(2), 193–204.
2. Choi, W.; Woo, M., *A Distributed Weighted Clustering Algorithm for Mobile ad Hoc Networks, in Proceedings of Advanced International Conference on Telecommunications and International Conference on Internet and Web Applications and Services* (AICT-ICIW'06), Guadeloupe, French Caribbean, 2006; pp 73–73.
3. Zabian, A.; Ibrahim, A.; Al-Kalani, F., Dynamic Head Cluster Election Algorithm for Clustered Ad-Hoc Networks. *J. Comput. Sci.* **2008**, 4(1), 42–50.
4. Chawla, M.; Singhai, J.; Rana, J. L. Clustering in Mobile ad Hoc Networks: A Review. *Int. J. Comput. Sci. Inf. Securi.* **2010**, 8(2), 293–301.
5. Mehta, S.; Sharma, P.; Kotecha, K., A Survey on Various Cluster Head Election Algorithms for MANET, in Proceedings of **2011** *Nirma University International Conference on Engineering (NUiCONE),* Ahmedabad: India, 2011, pp 1–6.

6. Kim, H. An Efficient Clustering Scheme for Data Aggregation Considering Mobility in Mobile Wireless Sensor Networks. *Int. J. Control Autom.* **2013**, *6*(1), 221–234.

7. Chatterjee, M.; Das, S. K.; Turgut, D. *A weight based distributed clustering algorithm for mobile ad hoc networks, in Proceedings of the 7th International Conference on High Performance Computing (HiPC2000)*, Bangalore, India, 2000; pp 511–521.

8. Agarwal, R.; Gupta, R.; Motwani, M., Review of Weighted Clustering Algorithms for Mobile ad Hoc Networks. *Comput. Sci. Telecommun.* **2012**, *33*(1), 71–78.

9. Sikander, G.; Zafar, M.; Raza, A.; Babar, M.; Mahmud, S.; Khan, G. A Survey of Cluster-Based Routing Schemes for Wireless Sensor Networks. *Smart Comput. Rev. Networks*, **2013**, *3*(4), 261–275.

10. Abbasi, A. A.; Younis, M. A Survey on Clustering Algorithms for Wireless Sensor Networks. *Comput. Commun.* **2007**, *30*(14), 2826–2841.

11. Darabkh, K. A.; Ismail, S. S.; Al-Shurman, M.; Jafar, I. F.; Alkhader, E.; Al-Mistarihi, M. F. Performance Evaluation of Selective and Adaptive Heads Clustering Algorithms Over Wireless Sensor Networks. *J. Network Comput. Appl.* **2012**, *35*(6), 2068–2080.

12. Geetha, V.; Kallapur, P. V.; Tellajeera, S. Clustering in Wireless Sensor Networks: Performance Comparison of LEACH & LEACH-C Protocols Using NS2. *Procedia. Technol.* **2012**, *4*:163–170

13. Wang, Y.; Wu, X.; Wang, J.; Liu, W.; Zheng, W. An OVSF Code Based Routing Protocol for Clustered Wireless Sensor Networks. *Int. J. Future Gener. Commun. Networking*, **2012**, *5*(3), 117–128.

14. Gajurel, S. Multi-Creteria Clustering, Case Western Reserve University Cleveland, OH. USA, Technical Report, 2006.

15. da Silva, A. P. R.; Martins, M. H.; Rocha, B. P.; Loureiro, A. A.; Ruiz, L. B.; Wong, H. C. *Decentralized Intrusion Detection in Wireless Sensor Networks, In Proceedings of the 1st ACM International Workshop on Quality of Service & Security in Wireless and Mobile Networks*, Montreal, Canada, 2005; pp 16–23.

16. Lehsaini, M.; Guyennet, H.; Feham, M. An Efficient Cluster-Based Self-Organisation Algorithm for Wireless Sensor Networks. *Int. J. Sens. Networks*, **2010**, *7*(1), 85–94.

17. Baker, D. J.; Ephremides, A. The Architectural Organization of Mobile Radio Network Via a Distributed Algorithm. *IEEE Trans. Commun.* **1981**, *29*(11), 1694–1701.

18. Gerla, M.; Tsai, J. T. C. Multi Cluster, Mobile, Multimedia Radio Network. *Wireless Networks* **1995**, *1*(3), 255–265.

CHAPTER 3

SECURITY IN WSNs

3.1 INTRODUCTION

Security and privacy are deemed as one of the challenges encountered in all types of wired and wireless networks. These challenges have great importance in wireless sensor networks (WSNs), where the unique characteristics of these networks and the application purposes they serve make them attractive targets for intrusions and other attacks.

In applications such as battlefield surveillance and assessment, target tracking and monitoring civil infrastructure such as bridges and tunnels, and assessment of disaster zones to guide emergency response activities, any breach of security, compromise of information, or disruption of correct application behavior can have very serious consequences. Sensor networks are frequently used in remote areas, left to operate unattended and therefore providing an easy target for physical attacks, unauthorized access, and tampering. Sensor nodes are typically very resource-constrained and operated in harsh environments, which further facilitate compromises and makes it often difficult to distinguish security breaches from node failures, varying link qualities, and other commonly found challenges in sensor networks. Finally, security mechanisms that are customized for WSNs applications are required for these resource constraints, such that the limited resources are used efficiently.[11] WSNs are susceptible to multiple types of attacks because they are randomly deployed in open and unprotected environments.[1,12,18,27,36] For securing WSNs, it is necessary to address the potential attacks on such networks. These can be classified as either passive attacks or active ones.[15,28] It is known that routing protocols in sensor networks are simpler and more vulnerable to attacks than the other two types of wireless networks: ad hoc and cellular. The first serious discussion and analyses on secure routing were performed by Karlof and Wagner.[20] They studied multiple types of attacks on routing protocols in detail and the effects on common routing protocols in WSNs. The assumption is that there are two types of attacks: outer and inner attacks.

Other researchers have used a decentralized approach to monitor network nodes with fault detections through the coordination of neighboring nodes[35] or the use of watchdogs to detect misbehavior in neighbors.[19,21,22,29] Da Silva et al.[36] adopted local monitoring between neighboring nodes. Among the studies that have been conducted in the related works, no research has intended to use a monitoring mechanism with a cluster-based architecture except scheme in Ref [2]. However, authors focused only on the misbehavior of malicious nodes and not on the nature of attacks. We thus propose a mechanism that assures the distributed monitoring of WSNs security issues. This mechanism uses a cluster-based architecture together with a new set of metrics and rules for diagnosing the state of the sensors. Reducing the flow of communication and providing a stable surveillance environment are the most significant advantages of this solution.

The intent of this chapter is to investigate the security-related issues in WSNs. First, the security architecture of sensor networks is proposed, trying to outline a general illustration in this area. Then, the following four aspects are investigated.[24] The cryptographic mechanisms

1. Various keying mechanisms for the key management issue
2. A panoramic view and detailed analysis of the trust management
3. A set of effective strategies based on protecting location privacy

3.2 SECURITY REQUIREMENTS AND RELATED ISSUES

A sensor network is considered a special type of network which shares some commonalities with atypical computer network. However, it poses unique requirements of its own. To that end, it is possible to say that the requirements of a WSN as encompassing both the typical network demands and the unique necessities suited solely to WSNs.[9]

3.2.1 DATA CONFIDENTIALITY

Data confidentiality is deemed as the most important issue in network security. Every network with any security focus will usually address this problem first. In sensor networks, the confidentiality relates to the following:[7,32]A sensor network should not leak sensor readings to its neighbors.

- In many applications, nodes communicate highly sensitive data, for example, key distribution; therefore, it is extremely necessary to build a secure channel in a WSN.
- Public sensor information, such as sensor identities and public keys, should also be encrypted to some extent to protect against traffic analysis attacks.
- The standard approach for keeping sensitive data secret is to encrypt the data with a secret key that only intended receivers possess, thus achieving confidentiality.

3.2.2 AUTHENTICATION

An adversary function is not only concerned with modifying the data packet. It is also able to change the whole packet stream by injecting additional packets. Therefore, it is essential for the receiver to ensure that the data used in any decision-making process originates from the correct source. On the other hand, while constructing the sensor network, authentication is necessary for many administrative tasks (e.g., network reprogramming or controlling sensor node duty cycle). From the above, it is possible to say that the authentication message is necessary for many applications in sensor networks. Informally, data authentication provides the opportunity to a receiver to verify whether the data is really sent by the claimed sender. In the case of two-party communication, data authentication can be achieved through a purely symmetric mechanism: the sender and the receiver share a secret key to compute the message authentication codes (MACs) of all communicated data. Perrig et al.[33] propose a key-chain distribution system for their μTESLA secure broadcast protocol. The basic idea of the μTESLA system is to achieve asymmetric cryptography by delaying the disclosure of the symmetric keys. One limitation of this system is that some initial information must be unicast to each sensor node before authentication of broadcast messages can begin. Liu and Ning[25,26] propose an enhancement to the μTESLA system that uses broadcasting of the key-chain commitments rather than μTESLA's unicasting technique.

3.2.3   DATA INTEGRITY

The implementation of confidentiality may unable an adversary to steal information. However, the adversary is able to change the data, so as to send the sensor network into disarray which means that the data is not safe. For example, a malicious node may add some fragments or manipulate the data within a packet. This new packet can then be sent to the original receiver. Because of the harsh communication, environment data loss or damage can even occur without the presence of a malicious node. Thus, data integrity ensures that any received data has not been altered in transit.[9]

3.2.4   DATA FRESHNESS

Inspite of assuring confidentiality and data integrity, it is crucial step to ensure the freshness of each message. Informally, data freshness suggests that the data is recent, and it ensures that no old messages have been replaced. This requirement is especially important when there are shared-key strategies employed in the design. Despite the fact that it takes time for new shared keys to be propagated to the entire network, they typically need to be changed over time.

In this case, it is easy for the adversary to use a replay attack, to disrupt the normal work of the sensor, if the sensor is unaware of the new key change time. To solve this problem, a nonce, or another time-related counter, can be added into the packet to ensure data freshness.[9]

3.2.5   AVAILABILITY

Some extra costs will be introduced by adjusting the traditional encryption algorithms to fit within the WSN. Modifying the code is chosen by some approaches as an option to reuse as much code as possible, whereas to achieve the same goal, some approaches try to make use of additional communication. What is more, some approaches force strict limitations on the data access or propose an unsuitable scheme (such as a central point scheme) in order to simplify the algorithm. But all these approaches weaken the availability of a sensor and sensor network for the following reasons:[9]Additional computation consumes additional energy. If no more energy exists, the data will no longer be available.

- Additional communication also consumes more energy. What is more, as communication increases, so it does the chance of incurring a communication conflict.
- If using the central point scheme, a single point of failure will be introduced. This greatly threatens the availability of the network. The requirement of security not only affects the operation of the network but also is highly important in maintaining the availability of the whole network.

3.2.6   SELF-ORGANIZATION

A WSN is typically an ad hoc network which requires every sensor node to be independent and flexible enough to be self-organizing and self-healing according to different situations. There is no fixed infrastructure available for the purpose of network management in a sensor network. This inherent feature brings a great challenge to WSN security as well. For example, the dynamics of the whole network inhibits the idea of preinstallation of a shared key between the base station and all sensors.[13] Several random key pre-distribution schemes have been proposed in the context of symmetric encryption techniques.[9,13,25] In the context of applying public-key cryptography (PKC) techniques in sensor networks, an efficient mechanism for public-key distribution is necessary as well. In the same way that distributed sensor networks must self-organize to support multi-hop routing, they must also self-organize to conduct key management and build trust relation among sensors. If self-organization is lacking in a sensor network, the damage resulting from an attack or even the hazardous environment may be devastating.

3.2.7   TIME SYNCHRONIZATION

Most sensor network applications rely on some form of time synchronization. An individual sensor's radio may be turned off for periods of time for the sake of conserving power. In addition, sensors may wish to compute the end-to-end delay of a packet as it travels between two pairwise sensors. A more collaborative sensor network may require group synchronization for tracking applications, and so forth. In ref [14], the authors propose a set of secure synchronization protocols for sender–receiver (pairwise), multi-hop sender–receiver (for use when the pair of nodes are not within single-hop range), and group synchronization.

3.2.8   SECURE LOCALIZATION

The utility of a sensor network will often depend on its ability to accurately and automatically locate each sensor in the network. The accurate location information is necessary for the designed sensor network to locate faults in order to pinpoint the location of a fault. Unfortunately, by reporting false signal strengths and replaying signals, it becomes easy for an attacker to manipulate nonsecured location information.

A technique called verifiable multilateration (VM) is described in ref [6]. In multilateration, from a series of known reference points, a device's position can be accurately computed. In ref [6], authenticated ranging and distance bounding are used to ensure accurate location of a node. Because of distance bounding, an attacking node can only increase its claimed distance from a reference point. However, to ensure location consistency, an attacking node would also have to prove that its distance from another reference point is shorter.[6] Since it cannot do this, a node manipulating the localization protocol can be found. For large sensor networks, the secure positioning for sensor networks algorithm is used. It is a three-phase algorithm based upon VM.[6] In ref [23], secure range-independent localization (SeRLoc) is described. Its novelty is its decentralized, range-independent nature. SeRLoc uses locators that transmit beacon information. It is assumed that the locators are trusted and cannot be compromised. Furthermore, each locator is assumed to know its own location. A sensor computes its location by listening for the beacon information sent by each locator.

The beacons include the locator's location. Using all of the beacons that a sensor node detects, a node computes an approximate location based on the coordinates of the locators. Using a majority vote scheme, the sensor then computes an overlapping antenna region. The final computed location is the "center of gravity" of the overlapping antenna region.[23] All beacons transmitted by the locators are with a shared global symmetric key that is preloaded to the sensor prior to deployment. Each sensor also shares a unique symmetric key with each locator. This key is also preloaded on each sensor.

3.3   MALICIOUS NODES ATTACKS IN WIRELESS SENSOR NETWORKS (WSNS)

WSNs are particularly vulnerable to a variety of security threats, such as malicious nodes on the transmission paths dropping, fabricating, or tampering the forwarded messages, and denial of service (DOS), while promoting a range of fundamental research challenges. The typical attacks in WSN include wormhole attack, sinkhole attack, Sybil attack, and so on, in which malicious nodes always try to participate in a path or compromise the nodes on path, so as to drop, fabricate, or tamper messages. There are many studies that have been carried out in these security threats. For example, Wood[40] classified attacks into different layers: each layer is susceptible to different attacks and has different options available for its defense. Some attacks crosscut multiple layers or exploit interactions between them.

3.3.1   PHYSICAL LAYER ATTACKS

Since the use of technology of wireless communication in WSN, it is easy to incur jamming attack from attackers in physical layer. Moreover, because of the placement of sensor nodes in an unguarded environment, physical access to the sensor node becomes possible. Therefore, an intruder may be able to tamper or damage with the sensor devices (jamming, tampering, and so forth).[9]

3.3.2 LINK LAYER ATTACKS AND COUNTERMEASURES

The link or medium access control (MAC) layer provides channel arbitration for neighbor-to-neighbor communication. Cooperative schemes that depend on carrier sense, which let nodes detect if other nodes are transmitting, are particularly vulnerable to all kinds of attacks. For example, collisions and unfairness at the link layer may be able to delay the packet transmission or cause the packet to be corrupted (collision, unfairness, exhaustion, and so forth).[9]

3.3.3 NETWORK AND ROUTING LAYER ATTACKS AND COUNTERMEASURES

Network layer attacks are a significant and credible threat to WSNs. This layer provides a critical service. Before reaching their destination, messages may pass through a lot of hops in a large-scale deployment. Unfortunately, as the aggregate network cost of relaying a packet increases, the probability of the dropping or misdirecting packet along the way in the network increases as well (homing, neglect and greed, misdirection, black holes, and so forth).[9]

3.3.4 TRANSPORT LAYER ATTACKS AND COUNTERMEASURES

Transport layer manages "end-to-end" connections and this layer is needed when the sensor network intends to be accessed through the Internet.

The service the layer provides can be as simple as an unreliable area-to-area any-cast, or as complex and costly as a reliable sequenced-multicast byte stream.

Sensor networks tend to use simple protocols to minimize the communication overhead of acknowledgments and retransmissions. The transport layer can be attacked through flooding or desynchronization (flooding, desynchronization, and so on).

3.4   SECURITY REQUIREMENTS

3.4.1   SECURITY GOALS

Various security requirements on sensor networks are presented in almost all the related papers.[24,32] These requirements can be classified into three levels:

3.4.1.1   MESSAGE-BASED LEVEL

Similar to that in conventional networks, this level deals with data confidentiality, authentication, integrity, and freshness. Symmetric-key cryptography and MACs are necessary security primitives to support information flow security. Moreover, data freshness is necessarily required as lots of content-correlative information is transmitted on a sensor network during a specific time.

3.4.1.2   NODE-BASED LEVEL

Situations such as node compromise or capture are investigated on this level. In case that a node is compromised, loaded secret information may be improperly used by adversaries.

3.4.1.3   NETWORK-BASED LEVEL

At this level, more network-related issues are addressed, as well as security itself. A major benefit of sensor networks is that they perform in-network processing to reduce large streams of raw data into useful aggregated information. Protecting it is critical. The security issue becomes more challenging when discussed seriously in specific network environments. First, securing a single sensor is completely different from securing the entire network, thus the network-based anti-intrusion abilities have to be estimated. Moreover, such network parameters as routing, node's energy consumption, signal range, network density, and so forth should be discussed correlatively. Moreover, the scalability issue is also important with respect to the redeployment of node addition and revocation.

3.4.2   *PERFORMANCE METRICS*

As addressed above, it is definitely insufficient to access a scheme based on its ability to provide secrecy. Ref [8] proposes the following evaluation metrics:

1. Resilience against node capture: On the network-based level, the fraction of total communications that are compromised is required to be estimated once a capture of several nodes occurs.
2. Resistance against node replication: This issue needs to be seriously investigated as the captured node may be cloned and thus adversaries gain more control of the network.
3. Revocation: Similat to regular process on node addition, the revocation mechanism is always necessary for detection and insulation of the misbehaving nodes.
4. Scale: Performance of the above security characteristics needs to be generally inspected, corresponding to different network scales.

3.5   HIERARCHICAL ARCHITECTURE FOR WSNS SECURITY

3.5.1   *THREE-LEVEL SECURITY REQUIREMENTS ARCHITECTURE ON SECURITY MECHANISMS*

For the sake of giving a general view on security issues addressed in sensor networks, the security architecture of sensor networks is presented in Figure 3.1. As it is described above, the three-level security requirements outline the principles of algorithm design on security mechanisms. The corresponding issues for each level in detail were listed so that securing available communication and applications in sensor networks are achieved, such as identity authentication, routing, data aggregation, and so forth.

There are three security research aspects most focused on: security primitives, key management, and network-related security strategies. Security primitives manage a minimal protection to information flow and a foundation to create secure protocols. These security primitives are as follows: systematical key encryption, MACs, and PKC. The issue of network-related security strategies is responsible for three main operations: combining communication throughout the entire network, integrating power and routing awareness, and promoting holistic working performance within tolerable costs.[4,10]

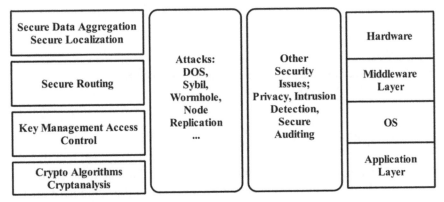

FIGURE 3.1   Security architecture of sensor networks.
Source: Adapted from ref [9].

3.5.2   SECURITY ARCHITECTURE (SECURITY MAP) OF SECURITY ISSUES IN WSNS

The new security architecture (security map) of security issues in WSNs is drawn as in Figure 3.2. Justifying and ensuring security is necessary before the large-scale deployment of sensors. The vertical comparison in Figure 3.2 demonstrates that various security issues are rendered in every layer of the protocol stacks from physical layer to applicable one. Despite the fact that it is hard to guarantee the security of every layer, it is possible to deal with the problems one by one and build appropriate security mechanisms satisfying particular appliances.[10]

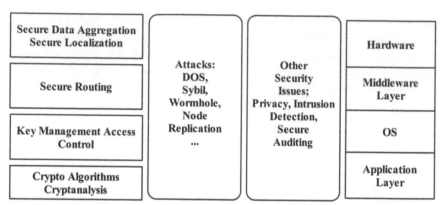

FIGURE 3.2   Security issues in wireless sensor networks (WSNs).
Source: Adapted from ref [9].

3.6   ATTACKS

An attack is a set of computer techniques, to cause damage to a network by exploiting flaws in it. The attacks can be located (on the same network) or remote (on the Internet or telecommunications). They can aggravate security problems. Indeed, the consequences of these attacks can vary from a simple traffic listening to the total shutdown of the network capacity as attackers. To fight them, it is necessary to know the classes and types of attacks to implement optimal solutions.[34] Attacks know several possible classifications; the most used are grouped into the following categories:[11]

3.6.1   DEPENDING ON THE DEGREE

- A class of attack mote (sensor) versus an attack of laptop class
  - Attack class mote: It is limited to a few nodes with similar capabilities to a single sensor. Thus, at some point, it can only monitor communications between a limited number of nodes.
  - Attack class laptop: Access to more powerful devices (sink), such as laptops. They have many more capabilities and a wide range of connectivity, the radius of their surveillance zone could monitor the entire network. This will give the opponent an advantage since the network of sensors can launch more serious attacks.

3.6.2   DEPENDING ON THE ORIGIN

- An external attack versus an internal attack
    - External attack: It is triggered by a node that does not belong to the network or does not have access permission.
    - Internal attack: It is triggered by a malicious internal node. Defence strategies generally aim to fight external attacks. However, internal attacks are the most serious threats that can disrupt the smooth functioning of WSNs.

3.6.3   DEPENDING ON THE NATURE

- A passive attack versus an active attack
    - Passive attack: It is triggered when a node obtains unauthorized access to a resource without modifying the data or disrupts network operation. Once the attacker has gained sufficient information, it can produce an attack against the network, which transforms the passive into an active attack.
    - Active attack: It is triggered when a node obtains unauthorized access to a resource by making changes to data or disrupting the smooth operation of the network.

3.7   OUR CLASSIFICATION

In this book, classification of attacks is based on:

1) Previous classification depending on the nature: passive attack versus an active attack.
2) The proposed classification by Stallings:[38] In such a classification, attacks can disrupt the normal flow of packets using the modification, interception, interruption, manufacturing, or combinations (cf. Fig. 3.3):
    - Interruption (attack against availability): A communication link is lost or unavailable.
    - Interception (attack against confidentiality): The network of sensors is compromised by an attacker who gains unauthorized access to a node or data exchanged through it.

- Modification (attack against integrity): The attacker made some changes to routing packets, and thus endangers its integrity in networks.
- Manufacturing (attack against authentication): The malicious node injects false information and undermines the reliability of the transmitted information.

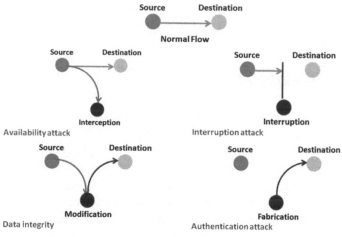

FIGURE 3.3   Classification of attacks according to Stallings.
Source: Adapted from ref [37].

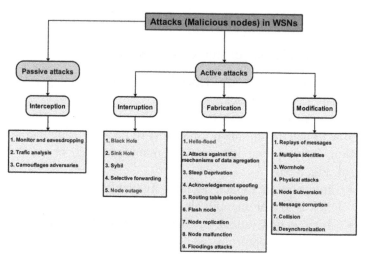

FIGURE 3.4   Our classification diagram attacks.

In this book, we only examine inner attacks and more precisely active attacks. Outer attacks are prevented by the use of link layer security mechanisms.[5]. In our current work, the focus is on the misbehavior of malicious nodes and the nature of attacks. In the following section, we review the most common network layer attacks on WSNs we selected and highlight the characteristics of these attacks[3,16,17,20] (cf. Fig. 3.4).

3.7.1   BLACK HOLE ATTACK

In this attack, malicious nodes advertise very short paths (sometimes zero-cost paths) to every other node, forming routing black holes within the network.[15] As their advertisement propagates, the network routes more traffic in their direction. In addition to disrupting traffic delivery, this causes intense resource contention around the malicious node as neighbors compete for limited bandwidth. These neighbors may themselves be exhausted prematurely, causing a hole or partition in the network.[10,39]Example in Figure 3.5:

The malicious node 5 stops transmitting packets sent by the nodes 3 and 7. As a result, the node 5 causes DOS to two nodes (3, 7).

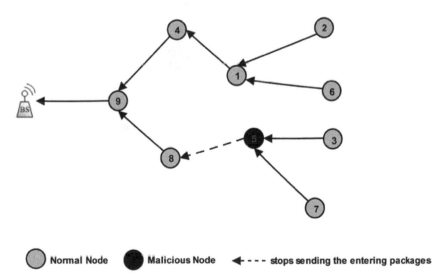

FIGURE 3.5   Black hole attack.

3.7.2 SINKHOLE ATTACK

Sinkhole attack is one of the most devastating ones: it is very hard to protect against.[18,38] In a sinkhole attack, the adversary's goal is to redirect nearly all the traffic from a particular area through a compromised node, creating a metaphorical sinkhole with the adversary at the center so that all traffic in the surrounding will be absorbed by the malicious node, because nodes, on or near the path followed by transmitted packets, have many opportunities to tamper with application data. Sinkhole attacks can enable many other attacks such as selective forwarding.[27]Example in Figure 3.6:

The attacker node 5 adverts to sink a single jump, compared with two real jumps. Consequently, nodes (1, 7) select node 5 to connect them to the sink, therefore as compared with the last attack, the attacker causes DOS to five nodes (1, 2, 3, 6,7).

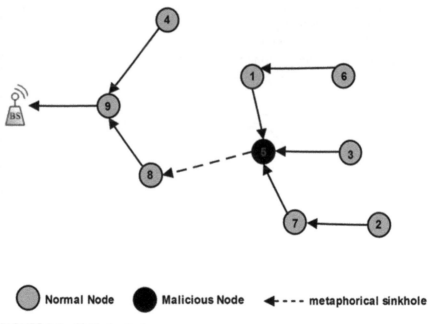

FIGURE 3.6   Sinkhole attack.

3.7.3 SELECTIVE FORWARDING ATTACK

In a selective forwarding attack, malicious nodes prevent the flow of routing information in sensor networks by refusing to forward or drop the messages traversing them.[10,20] Another aspect of this type of attack is that malicious nodes may forward the messages along an incorrect path, creating inaccurate routing information in the network.

Example in Figure 3.7.

The attacker node 5 transmits all packets except those it receives from node 4, based on the source address; it causes DOS to node 4 only, while remaining normal for all other nodes connected. A naive user can determine that node 4 is defective.

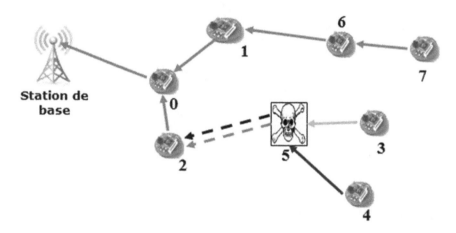

FIGURE 3.7   Selective forwarding.

3.7.4 NODE OUTAGE ATTACK

If a node acts as an intermediary, an aggregation point, or a cluster head, what happens if the node stops working?

Protocols used by the WSNs must be robust enough to mitigate the effects of failures by providing alternate routes [10,30,31]

3.7.5 HELLO FLOODS ATTACK

Many routing protocols use "Hello" broadcast messages to announce themselves to their neighbor nodes. The nodes that receive this message assume that source nodes are within range and add source nodes to their neighbor list.

The hello flood attacks can be caused by a node which broadcasts a Hello packet with very high power, so that a large number of nodes even far away in the network choose it as the parent node.[28] These nodes are then convinced that the attacker node is their neighbor, so that all the nodes will respond to the "Hello" message and waste their energy (cf. Fig. 3.8).

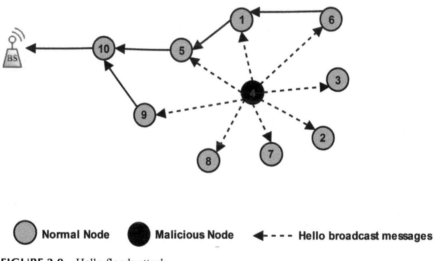

FIGURE 3.8 Hello floods attack.

3.7.6 SYBIL ATTACK

In this attack, a malicious node can present multiple identities to other nodes in the network. The Sybil attack poses a significant threat to most geographic routing protocols. Sybil attacks are prevented through link layer authentication (cf. Figs. 3.9 and 3.10).

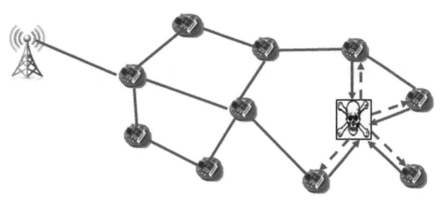

FIGURE 3.9   Sybil attack (the attack of multiple identities).

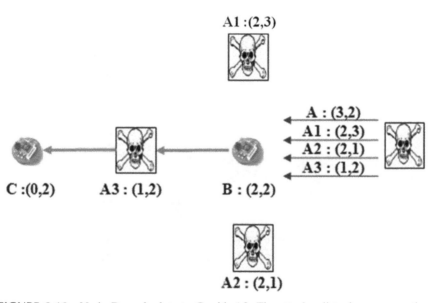

FIGURE 3.10   Node B sends data to C with A3, The attacker listening conversation Attacker A (3.2) collects several identities A1, A2, A3.

The Sybil attack poses significant threats:

- Threat to geographic routing:
  - Being in more than one place at once
- Threat to aggregation processing:
  - Sending multiple (fictitious) results to a parent
  - Sending data to more than one parent

3.7.7 WORMHOLE ATTACK

In a wormhole attack, the adversary tunnels a message received in one malicious node and replays them in a different part of the network (cf. Fig. 3.11). The two malicious nodes usually claim that they are merely two hops from the base station. Khalil et al.[21] suggests five modes of wormhole attacks in his paper. Details of these modes are in refs 21, and 37. Summarizing, the different modes of the wormhole attack along with the associated requirements are given in ref Table 3.1.

TABLE 3.1 Summary of Wormhole Attack Modes.

Mode name	Minimum of compromised nodes	Special requirements
Packet encapsulation	Two	None
Out-of-band channel	Two	Out-of-band link
High-power transmission	One	High-energy source
Packet relay	One	None
Protocol deviations	One	None

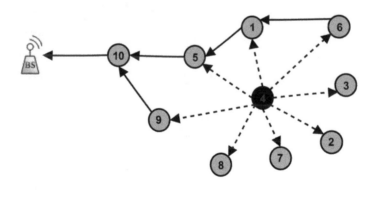

FIGURE 3.11 Wormhole through out-of-band channel.

3.8   CONCLUSION

Similar to every other computer network, WSNs are exposed to a variety of threats and attacks and as most other networks, sensor networks require support for confidentiality, integrity, and authentication to protect sensor nodes and sensor data.

However, several unique characteristics of WSNs, such as remote deployment (which facilitates an adversary's physical access to sensor nodes) and resource constraints make it easier to compromise sensors and sensor data. Further, many sensor networks are attractive targets for attackers due to the nature of many WSNs applications and the sensitive data they generate (e.g., military applications, emergency response, and health care).

This chapter provided a brief overview of several types of attacks commonly found in sensor networks and techniques and protocols to defend a network or to detect an intrusion or compromised node. As WSNs continue to become more commonplace, it is to expect that security challenges will increase, the types and number of threats will evolve, and new solutions to protect sensor networks and sensor data will be required.[11] In the next chapter, we propose our approach which focuses around strategy of distributed resolution which enables us to generate a reduced number of balanced and homogeneous clusters in order to minimize the energy consumption of the entire network and prolong sensors lifetime. The introduction of a new metric (the behavioral-level metric) promotes a safe choice of a cluster head in the sense where this last one will never be a malicious node. Thus, the highlight of our work is summarized in a comprehensive strategy for monitoring the network in order to detect and remove the malicious nodes.

KEYWORDS

- wireless sensor networks
- data confidentiality
- message authentication codes
- secure range-independent localization
- denial of service

REFERENCES

1. Anderson, R. J.; Kuhn, M. G. Low Cost Attacks on Tamper Resistant Devices. *Proceedings of the 5th International Workshop on Security Protocols*, 1998, 125–136.
2. Benahmed, K.; Merabti, M.; Haffaf, H.Distributed Monitoring for Misbehaviour Detection in Wireless Sensor Networks. *Secur. Commun.Netw.* **2013**,*6*(4), 388–400.
3. Berwal, P.Security in Wireless Sensor Networks: Issues and Challenges. *Int. J. Eng. Innov. Technol.* **2013**, *3*(5), 192–198.
4. Bettstetter, C.On the Minimum Node Degree and Connectivity of a Wireless Multihop Network. *Proceedings of the 3rd ACM International Symposium on Mobile Ad Hoc Networking and Computing (MobiHoc'02)*, 2002.
5. Camtepe, S. A.; Yener, B. *Key Distribution Mechanisms for Wireless Sensor Networks: A Survey*; Technical Report 05–07, Rensselaer Polytechnic Institute: Troy, New York, 2005.
6. Capkun, S.; Hubaux, J. Secure Positioning in Wireless Networks. *Proc. IEEE J. Sel. Areas Commun. Spec. Issue Secur. Wirel.Ad Hoc Netw.* **2006**, *24*(2), 221–232.
7. Carman, D.; Kruus, P.; Matt, B. J. *Constraints and Approaches for Distributed Sensor Network Security*; Technical Report 00–010; NAI Labs, 2000.
8. Chan, H.; Perrig, A.; Song, D. Random Key Predistribution Schemes for Sensor Networks. *Proceeding of the IEEE Symp.On Research in Security and Privacy. IEEE Comput. Soci.* **2003**,197–213.
9. Conti, M. Secure Wireless Sensor Networks Threats and Solutions. In *Advances in Information Security*; (Springer), 2016; Vol. 65, ISSN 1568–2633.
10. Dahane, A.; Berrached, N.; Loukil, A.Safety of Mobile Wireless Sensor Networks Based on Clustering Algorithm. *Int. J. Wirel.Netw.Broadband Technol. (IJWNBT)* **2016**, *5*(1), 72–100, (June).
11. Dargie, W.; Poellabauer, C. *Fundamentals of Wireless Sensor Networks: Theory and Practices: Wiley Series on Wireless Communications and Mobile Computing*;Wiley,2010.
12. Djamel, D. Security in Mobile Ad hoc Networks: Detection and Isolation of Selfish Nodes (Thèse de doctorat),2007.
13. Eschenauer, L.; Gligor, V. D.A Key-Management Scheme for Distributed Sensor Networks. *Proceedings of the 9th ACM Conference on Computer and Communications Security (CCS'02)*, 2002, 41–47.
14. Ganeriwal, S.; Srivastava, M. B. Reputation-Based Framework for High Integrity Sensor Networks. *Proceedings of the 2nd ACM Workshop on Security of Ad Hoc and Sensor Network (SASN2004)*,Washington, DC, 2004;pp66–77.
15. Ghamgin, H.; Akhgar, M. S.; Jafari, M. T.Attacks in Wireless Sensor Network. *J. Appl. Sci. Res.* **2011**,*7*(7), 954–960.
16. Hai, T. H.; Huh, E. N.; Jo, M.A Lightweight Intrusion Detection Framework for Wireless Sensor Networks.*Wirel.Commun.Mob.Comput.***2010**,10(4), 559–572.
17. Huh, E. N.; Hai, T. H. *Lightweight Intrusion Detection for Wireless Sensor Networks, Rijeka*; INTECH Open Access Publisher: Croatia, 2011.

18. Islam, M. S.; Rahman, S. A.Anomaly Intrusion Detection System in Wireless Sensor Networks: Security Threats and Existing Approaches. *Int. J. Adv. Sci. Technol.***2011,***36*(1), 1–8.

19. Kaplantzis, S.; Shilton, A.; Mani, N.; Sekercioglu, Y. A. Detecting Selective Forwarding Attacks in Wireless Sensor Networks Using Support Vector Machines. *Proceedings of 3rd International Conference on Intelligent Sensors, Sensor Networks and Information (ISSNIP2007)*, Melbourne, Australia, 2007;pp335–340.

20. Karlof, C.; Wagner, D.Secure Routing in Wireless Sensor Networks: Attacks and Countermeasures. *Ad Hoc Netw.***2003**, *1*(2);pp293–315.

21. Khalil, I.; Bagchi, S.; Shroff, N. B. LITEWORP: A Lightweight Countermeasure for the Wormhole Attack in Multihop Wireless Networks. *Proceedings of International Conference on Dependable Systems and Networks (DSN2005)*,Yokohama, Japan, 2005;pp612–621.

22. Khalil, I.; Bagchi, S.; Shroff, N. F.Mobiworp: Mitigation of the Wormhole Attack in Mobile Multihop Wireless Networks. *Ad Hoc Netw.***2008**, *6*(3), 344–362.

23. Lazos, L.; Poovendran, R.Serloc: Robust Localization for Wireless Sensor Networks. *ACM Trans. Sens. Netw.***2005**,*1*(1), 73–100.

24. Li, P.; Sun, L.; Fu, X.; Ning, L. *Security in Wireless Sensor Networks*; Chapter book, (Springer), 2013;pp179–227.

25. Liu, D.; Ning, P. Efficient Distribution of Key Chain Commitments for Broadcast Authentication in Distributed Sensor Networks. *Proceedings of the 10th Network and Distributed System Security Symposium (NDSS'03)*, 2003;pp263–276.

26. Liu, D.; Ning, P.Multilevelμtesla Broadcast Authentication for Distributed Sensor Networks. *ACM Trans. Embed Comput. Syst.* **2004,***3*(4), 800–836. http://doi.acm.org/10.1145/1027794.1027800.

27. Marchesani, S.; Pomante, L.; Pugliese, M.; Santucci, F.A Middleware Approach to Provide Security in IEEE 802.15.4, Wireless Sensor Networks. *Proceedings of International Conference on Mobile Wireless Middleware, Operating Systems and Applications (Mobilware)*, Bologna, Italy, 2013;pp85–93.

28. Martins, D.; Guyennet, H.Security in Wireless Sensor Networks: A Survey of Attacks and Countermeasures. *Int. J. SpaceBased Situated Comput.***2011,***1*(2), 151–162.

29. Marti, S.; Giuli, T. J.; Lai, K.; Baker, M. Mitigating Routing Misbehavior in Mobile Ad Hoc Networks. *Proceedings of the 6th Annual International Conference on Mobile Computing and Networking (MobiCom)*, Boston, MA, 2005;pp255–265.

30. Padmavathi, D. G.; Shanmugapriya, D. A Survey of Attacks, Security Mechanisms and Challenges in Wireless Sensor Networks. *Int. J. Comput. Sci. Inf. Secur.***2009,***4*(1–2), 1–9.

31. Pathan, A. S. K.; Lee, H. W.; Hong, C. S., Security in Wireless Sensor Networks: Issues and Challenges. *Proceedings of the 8th International Conference Advanced Communication Technology (ICACT2006)*, Phoenix Park, Korea, 2006;pp1048–1054.

32. Perrig, A.; Szewczyk, R.; Wen, V.; Culler, D. E.; Tygar, J. D. SPINS: Security Protocols for Sensor Networks. *Proceedings of the 7th Annual ACM/IEEE International Conference on Mobile Computing and Networking (MobiCom'01)*, 2001;pp189–199.

33. Perrig, A.; Szewczyk, R.; Tygar, J. D.; Wen, V.; Culler, D. E.Spins: Security Protocols for Sensor Networks. *Wirel.Netw.* **2002,***8*(5), 521–534.

34. Safiqul-Islam, M.; Ashiqur-Rahman, S.Anomaly Intrusion Detection System in Wireless Sensor Networks: Security Threats and Existing Approaches. *Int. J. Adv. Sci. Technol.***2011**, *36*,1–8.

35. Sheth, A.; Hartung, C.; Han, R.A Decentralized Fault Diagnosis System for Wireless Sensor Networks. *Proceedings of IEEE International Conference on Mobile Ad-hoc and Sensor Systems Conference,*Washington, DC, 2005.

36. da Silva, A. P. R.; Martins, M. H.; Rocha, B. P.; Loureiro, A. A.; Ruiz, L. B.; Wong, H. C. Decentralized Intrusion Detection in Wireless Sensor networks. *Proceedings of the 1st ACM International Workshop on Quality of Service and Security in Wireless and Mobile Networks*, Montreal, Canada, 2005,pp16–23.

37. Stallings, W. *Cryptography and Network Security: Principles and Practices*, 5th ed.; Pearson Education: Harlow,2010.

38. Taneja, S.; Kush, A. A Survey of Routing Protocols in Mobile Ad Hoc Networks. *Int. J. Innovation, Manage. Technol.* **2010**, *1*(3), 279–285.

39. Tripathi, M.; Gaur, M. S.; Laxmi, V.(June)Comparing the Impact of Black Hole and Gray Hole Attack on Leach in Wsn. *Proceedings of the 4th International Conference on Ambient Systems. Networks Technologies (ANT '13) the 3rd International Conference on Sustainable Energy Information Technology (SEIT '13),*2013, 19,1101–1107.

40. Wood, A. D.; Stankovic, J. A.Denial of Service in Sensor Networks. *Computer* **2002,***35*(10), 54–62.

CHAPTER 4

WSNs ROUTING PROTOCOLS

4.1 INTRODUCTION

Communication in wireless sensor network (WSN) is subject to various phenomena that characterize the communication by radio wave. The most commonly known is the signal attenuation with distance, which prevents the communication between two nodes that are without direct communication with each other and forces the packets relay through-intermediate nodes. The same can be said for message corruption when two close nodes emit simultaneously (collision), and problems like hidden and exposed nodes, which are phenomena proper to WSNs. Therefore, the routing protocol must be efficient to overcome these problems.

The first routing protocols were designed for the mobile ad hoc networks (MANET),where the bandwidth, the delay, and the mobility are the main concerns. Nevertheless, these protocols proved to be inefficient for WSNs because they are not optimized for the networks where the energy is a priority. Therefore, the developed protocols must ensure minimum energy consumption while maintaining the proper functioning of the network without degrading its performance.

In this chapter, we present the main routing metrics and the protocols criteria in their classification and citing some examples of protocols, with detailed description for the ad hoc on demand distance vector and destination sequenceddistance vector protocols used for the simulation.

4.2 ROUTING METRICS

This section examines the most common metrics used to measure the effectiveness of routing protocols.The routing metrics place a more important part in routingprotocols' design and performance; every routing metric indicates the cost ofpath selection and the routing capacity.[32] Routing protocols allow nodes tocompare calculated metrics to determine the optimal routes to borrow.Several metrics can affect the routing in terms of energy, time, pathlength, and so forth. In addition, they can be considered alone or combined (as hybridmetrics).[9]

4.2.1 METRICS FOR ENERGY CONSUMPTION

Routing protocols use this set of metrics to minimize energy consumptionduring routing[8]. The idea is to calculate the available energy (AE) for each node and the required energy (RE) for packets transmission between two nodes.[6] The route between the nodes and the sink isestablished so that each oneis characterized by the sum of nodes AE that constitute it and by the sum oflinks RE which build it. Energy consumption follows several approaches whichinclude:

- In consideration of power

The chosen route is characterized by the greater sum of AE.

- In consideration of cost
  The chosen route is characterized by the smallest amount of RE.
- In consideration of power and cost

This metric is a combination of the two previous metrics. The chosen route is characterized by the smallest sum of RE and the greatest amount of AE.

4.2.2 NUMBER OF HOPS

Routing protocols use this metric to minimize the number of hops during routing. The idea is to calculate the number of intermediate nodes that can be traversed during the transmission of a packet from source node to the sink. The route chosen contains the least number of nodes (minimum hops).[25]

4.2.3 PACKETS LOSS

Routing protocols use this metric in order to minimize the number of data packets lost during the transfer from a source to a destination during routing. The idea is to calculate the ratio of lost and transmitted packets transiting the network. In other words, we divide the number of lost packets by the number of transmitted packets during a transmission.[6] In the case where the packet loss rate is high, it is necessary to establish mechanisms to solve this problem.

4.2.4 END-TO-END DELAY (EED)

End-to-end delay (EED)is the sum of delays experienced at each hop from the source to the destination. The delay at each intermediate node has two components: a fixed delay, which includes the transmission at sender node and the propagation over the link to the next node, and a variable delay, which includes the processing and queuing at the sender node.[17]

This technique is among the best known metrics in wireless networks. Routing protocols use it to minimize the propagation delay of exchanged data packets during routing.

4.3 ROUTING PROTOCOLS CLASSIFICATION

Recently, routing protocols designed for WSN have been widely studied. The methods can be classified according to several criteria which are illustrated in Figure 4.1.

This classification helps the user to choose the appropriate infrastructure for one'sapplication.[1]

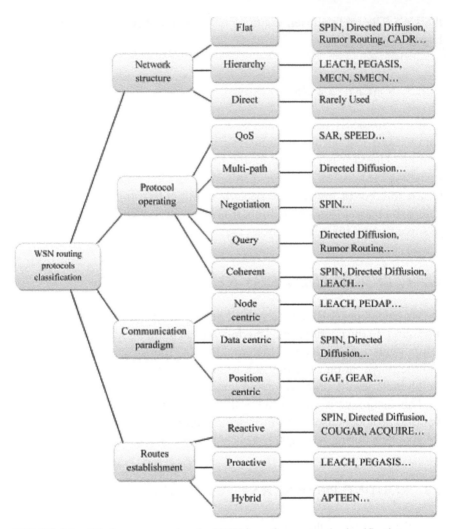

FIGURE 4.1   Wireless sensor networks (WSNs) routing protocols classification.

4.3.1 *NETWORK STRUCTURE-BASED PROTOCOLS*

The underlying network structure can play a significant role in the routing protocol operation in WSNs. In this section, we survey in detail most of the protocols that fall below this category.

4.3.1.1 FLAT ROUTING

The first category of routing protocols is the multi-hop flat routing protocols. In flat networks, each node typically plays the same role and sensor nodes collaborate together to perform the sensing task.

4.3.1.1.1   Flooding and Gossiping

Flooding and gossiping are two classical mechanisms to relay data in sensor networks without the need for any routing algorithms and topology maintenance. In flooding, each sensor node will transfer those messages received to all the neighbor nodes, and this process will be repeated until the messages arrive at sink node or is overtime due Time-To-Live(TTL,usually defined as the largest hop in WSNs). Gossiping improves flooding algorithm in some ways, and each sensor node only transfer the messages to a random neighbor node. However, even though flooding and gossiping is very simple and suitable for any network structure, both algorithms are not practical in application specified network, and they can easily bring implosion and overlap problems (see Fig. 4.2 and Fig. 4.3).

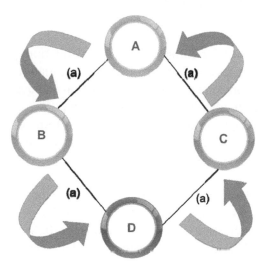

FIGURE 4.2   The implosion problem. Node A floods its data to all of its neighbors. D gets two same copies of data eventually, which is not necessary.
Source: Adapted from ref [15].

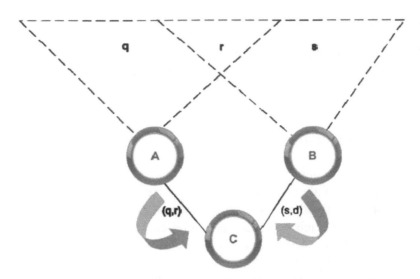

FIGURE 4.3   The overlap problem. Two sensors cover an overlapping geographic region and C gets same copy of data from these sensors.
Source: Adapted from ref [15].

4.3.1.1.2   Sensor Protocol for Information via Negotiation (SPIN)

The Sensor Protocol for Information via Negotiation (SPIN) family of protocols incorporates two key innovations: negotiation and resourceadaptation. To overcome the problems of implosion and overlap, SPIN nodes negotiate with each other before transmitting data.

Negotiation helps ensure that only useful information will be transferred. To negotiate successfully, nodes must be able to describe or name the data they observe. We refer to the descriptors used in SPIN negotiations as metadata.

In SPIN, nodes poll their resources before data transmission. Each sensor node has its own resource manager that keeps track of resource consumption; applications probe the manager before transmitting or processing data. This allows sensors to cut back on certain activities when energy is low.[14]

Together, these features overcome the deficiencies of classic flooding. The metadata negotiation process that precedes actual data transmission eliminates implosion because it eliminates transmission of redundant data messages. The use of metadata descriptors eliminates the possibility of overlap because it allows nodes to name the portion of the data that they are interested in obtaining. Being aware of local energy resources allows sensors to cut back on activities whenever their energy resources are low, thereby extending longevity.[14]

The SPIN uses three primitives:

- ADV (ADVertise): When a SPIN node has some new data, it sends an ADV message to its neighbors containing metadata (data descriptor).
- REQ (REQuest): When a SPIN node wishes to receive the data, it sends an REQ message.
- DATA: These are actual data messages with a metadata header.

A node wishing to send a data begins by sending ADV packet. Nodes that receive this packet verify that they have not yet responded to this request; in this case, they return a REQ packet. The node that initiates the communication, sends a DATA packet to each received REQ (Fig. 4.4).[15]

One of the advantages of SPIN is that topological changes are localized since each node needs to know only its single-hop neighbors. SPIN gives a factor of 3.5 less than flooding in terms of energy dissipation, and metadata negotiation almost halves the redundant data. However, SPIN's data advertisement mechanism cannot guarantee the delivery of data. For instance, if the nodes that are interested in the data are far away from the source node and the nodes between source and destination are not interested in that data, such data will not be delivered to the destination at all. Therefore, SPIN is not a good choice for the applications such as intrusion detection, which require reliable delivery of data packets over regular intervals.

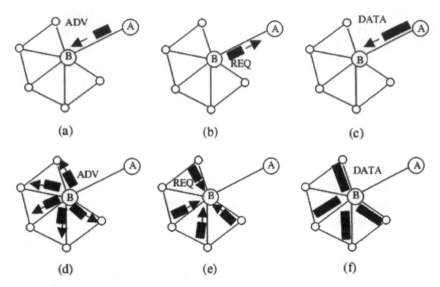

FIGURE 4.4   Sensor Protocol for Information through Negotiation (SPIN)protocol. Node A starts by advertising its data to node B (a). Node B responds by sending a request to node A (b). After receiving the requested data (c), node B sends out advertisements to its neighbors (d), who in turn send requests back to B (e–f).

Source: Ref [15] with permission.

4.3.1.1.3   *Directed Diffusion*

Directed Diffusion suggests the use of attribute–value pairs for the data and queries the sensors in an on-demand basis by using those pairs. In order to create a query, an interest is defined using a list of attribute–value pairs such as name of objects, interval, duration, geographical area, and so forth. The interest is broadcast by a sink through its neighbors. Each node receiving the interest can do caching for later use. The nodes also have the ability to do in-network data aggregation in order to reduce communication costs. The interest entry also contains several gradient fields. The interests in the caches are then used to compare the received data with the values in the interests (attribute–value). The interest entry also contains several gradient fields.

A gradient is a reply link to a neighbor from which the interest was received (link between node which sends the interest and the neighbor that replies to the interest). It is characterized by the data rate, duration, and expiration time derived from the received interest's fields. Hence, by utilizing interest and gradients, the paths are established between sink and sources.[1] Several paths can be established so that one of them is selected by reinforcement, Figure 4.5[15] summarizes the Directed Diffusion protocol.

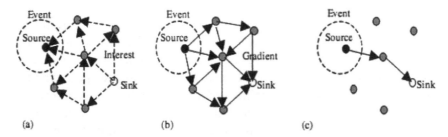

FIGURE 4.5   Directed Diffusion protocol phases. (a) Interest propagation, (b) initial gradients setup, (c) data delivery along reinforced.
Source: Ref [15] with permission.

Directed Diffusion differs from SPIN in terms of the on-demand data querying mechanism it has. In Directed Diffusion, the sink queries the sensor nodes if a specific data is available by flooding some tasks. In SPIN, sensors advertise the availability of data allowing interested nodes to query that data. Directed Diffusion has many advantages. Since it is data centric, all communication is neighbor-to-neighbor with no need for a node-addressing mechanism.

Each node can do aggregation and caching, in addition to sensing. Caching is a big advantage in terms of energy efficiency and delay. In addition, Direct Diffusion is highly energy efficient since it is on demand and there is no need for maintaining global network topology. However, Directed Diffusion cannot be applied to all sensor network applications since it is based on a query-driven data delivery model. The applications that require continuous data delivery to the sink will not work efficiently with a query-driven on-demand data model. Therefore, Directed Diffusion is not a good choice as a routing protocol for the applications such as environmental monitoring.[1]

4.3.1.1.4   Energy-Aware Routing

The potential problem in some protocols is that they find the lowest energy route and use that for every communication. However, that is not the best thing to do for network lifetime. Using a low-energy path frequently leads to energy depletion of the nodes along that path and in the worst case may lead to network partition.[24]

The objective of energy-aware routing protocol, a destination-initiated reactive protocol, is to increase the network lifetime. Although this protocol is similar to Directed Diffusion, it differs in the sense that it maintains a set of paths instead of maintaining or enforcing one optimal path at higher rates. These paths are maintained and chosen by means of a certain probability. The value of this probability depends on how low the energy consumption of each path can be achieved. By having paths chosen at different times, the energy of any single path will not deplete quickly. This can achieve longer network lifetime as energy is dissipated more equally among all the nodes. Network survivability is the main metric of this protocol. The protocol initiates a connection through localized flooding, which is used to discover all routes between source/destination pair and their costs; thus building up the routing tables.[3]

The high-cost paths are discarded and a forwarding table is built by choosing neighboring nodes in a manner that is proportional to their cost. Then, each node forwards the packet by randomly choosing a node from its forwarding table using the probabilities.

Localized flooding is performed by the destination node to keep the paths alive. When compared to Directed Diffusion, this protocol provides an overall improvement of 21.5% energy saving and a 44% increase in network lifetime. However, the approach requires gathering the location information and setting up the addressing mechanism for the nodes, which complicate route setup compared to the Directed Diffusion.[3]

4.3.1.1.5   Rumor Routing

It is a variation of Directed Diffusion and is mainly intended for applications where geographic routing is not feasible. In general, Directed Diffusion uses flooding to inject the query to the entire network when there is no geographic criterion to diffuse tasks. However, in some cases, there is only a little amount of data requested from the nodes and thus the use of flooding is unnecessary. An alternative approach is to flood the events if the number of events is small and the number of queries is large. The key idea is to route the queries to the nodes that have observed a particular event rather than flooding the entire network to retrieve information about the occurring events. In order to flood events through the network, the rumor routing algorithm employs long-lived packets, called agents. When a node detects an event, it adds such event to its local table called events table, and generates an agent.[7]

The agents travel the network in order to propagate information about local events to distant nodes. When a node generates a query for an event, the nodes that know the route may respond to the query by inspecting its event table. Hence, there is no need to flood the whole network, which reduces the communication cost. On the other hand, rumor routing maintains only one path between source and destination as opposed to Directed Diffusion,where data can be routed through multiple paths at low rates.[7]

Simulation results have shown (see Fig. 4.6) that rumor routing achieves significant energy saving over event flooding and can also handle node's failure.[1]

However, rumor routing performs well only when the number of events is small. For large number of events, the cost of maintaining agents and event tables in each node may not be amortized if there is not enough interest on those events from the sink.[1] Moreover, the overhead associated with rumor routing is controlled by different parameters used in the algorithm such as TTL pertaining to queries and agents.[7]

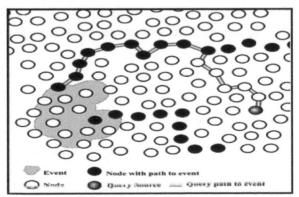

FIGURE 4.6   Rumor routing. Query is originated from the query source and searches for a path to the event. As soon as it finds a node on the path, it is routed directly to the event. *Source:* Ref [15] with permission.

4.3.1.1.6   Gradient-Based Routing (GBR)

Schurgers and Srivastava proposed another variant of Directed Diffusion, called Gradient-Based Routing (GBR). The key idea in GBR is to memorize the number of hops when the interest is diffused through the whole network. As such, each node can calculate a parameter called the height of the node, which is the minimum number of hops to reach the sink. The difference between a node's height and that of its neighbor is considered the gradient on that link. A packet is forwarded on a link with the largest gradient.[3]

The authors aim to usesome auxiliary techniques such as data aggregation and traffic spreading along with GBR in order to divide the traffic uniformly over the network. Nodes acting as a relay for multiple paths can create a data combining entity in order to aggregate data. On the other hand, three different data dissemination techniques have been presented:[1]

- Stochastic scheme: When there are two or more next hops with the same gradient, the node chooses one of them at random.
- Energy-based scheme: When a node's energy drops below a certain threshold, it increases its height so that other sensors are discouraged from sending data to that node.
- Stream-based scheme: The idea is to divert new streams away from nodes that are currently part of the path of other streams.

The main objective of these schemes is to obtain a balanced distribution of the traffic in the network, thus increasing the network lifetime. Simulation results of GBR showed that GBR outperforms Directed Diffusion in terms of total communication energy.

4.3.1.1.7   Constrained Anisotropic Diffusion Routing and Information-Driven Sensor Querying

Constrained Anisotropic Diffusion Routing (CADR) and Information-Driven Sensor Querying (IDSQ) are two routing techniques. CADR aims to be a general form of Directed Diffusion. The key idea is to query sensors and route data in the network such that the information gain is maximized,whereaslatency and bandwidth are minimized. CADR diffuses queries by using a set of information criteria to select which sensors can get the data. This is achieved by activating only the sensors that are close to a particular event and dynamically adjusting data routes. The main difference from Directed Diffusion is the consideration of information gain in addition to the communication cost. In CADR, each node evaluates an information/cost objective and routes data based on the local information/cost gradient and enduser requirements.[10]

In IDSQ, the querying node can determine which node can provide the most useful information with the additional advantage of balancing the energy cost. However, IDSQ does not specifically define how the query and the information are routed between the sensors and the sink. Therefore, IDSQ can be seen as a complementary optimization procedure. Simulation results showed that these approaches are more energyefficient than Directed Diffusion,where queries are diffused in an isotropic fashion and reaching nearest neighbors first.[10]

4.3.1.1.8   COUGAR

COUGAR views the network as a huge distributed database system. The key idea is to use declarative queries in order to abstract query processing from the network layer functions such as selection of relevant sensors and so on. COUGAR utilizes in-network data aggregation to obtain more energy savings.

The abstraction is supported through an additional query layer that lies between the network and application layers.[3]

COUGAR incorporates architecture for the sensor database system where sensor nodes select a leader node to perform aggregation and transmit the data to external base station (BS). The BS is responsible for generating a query plan, which specifies the necessary information about the data flow and in-network computation for the incoming query and send it to the relevant nodes. The query plan also describes how to select a leader for the query. The architecture provides in-network computation ability that can provide energy efficiency in situations when the generated data is huge.[3]

COUGAR providesnetwork-layer independent methods for data query. However, COUGAR has some drawbacks. First, the addition of query layer on each sensor node may add an extra overhead in terms of energy consumption and memory storage. Second, to obtain successful in-network data computation, synchronization among nodes is required (not all data are received at the same time from incoming sources) before sending the data to the leader node. Third, the leader nodes should be dynamically maintained to prevent them from being hotspots (failure prone).[3]

4.3.1.1.9   ACQUIRE

Similar to COUGAR, ACQUIRE (Active Query forwarding in sensor networks) views the network as a distributed database where complex queries can be further divided into several sub-queries. The operation of ACQUIRE can be described as follows. The base station (BS) node sends a query, which is then forwarded by each node receiving the query. During this, each node tries to respond to the query partially by using its pre-cached information and then forward it to another sensor node. If the pre-cached information is not up-to-date, the nodes gather information from their neighbors within a look-ahead of d hops.[23]

Once the query is being resolved completely, it is sent back through either the reverse or the shortestpath to the sink. Hence, ACQUIRE can deal with complex queries by allowing many nodes to send responses. Note that Directed Diffusion may not be used for complex queries due to energy considerations as Directed Diffusion also uses flooding-based query mechanism for continuous and aggregate queries. On the other hand, ACQUIRE can provide efficient querying by adjusting the value of the look-ahead parameter d. When d is equal to network diameter, ACQUIRE mechanism behaves similar to flooding. However, the query has to travel more hops if d is too small.[23]

A mathematical modeling was used to find an optimal value of the parameter d for a grid of sensors where each node has four immediate neighbors. However, there is no validation of results through simulation. To select the next node for forwarding the query, ACQUIRE either picks it randomly or the selection is based on maximum potential of query satisfaction.

Recall that selection of next node is based on either information gain (CADR and IDSQ) or query is forwarded to a node, which knows the path to the searched event (rumor routing).[23]

In this scheme (see Fig. 4.7), they use a one-hop look-ahead (d=1). At each step of the active query propagation, the node carrying the active query employs knowledge gained due to the triggered updates from all nodes within d hops in order to partially resolve the query. As d becomes larger, the active query has to travel fewer steps on average, andthis also raises the update costs. When d becomes extremely large, ACQUIRE starts to resemble traditional flooding-based querying.[23]

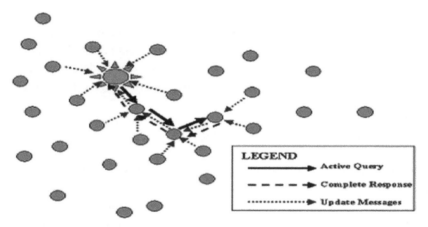

FIGURE 4.7   Illustration of ACQUIRE.
Source: Ref [15] with permission.

4.3.1.1.10   Minimum Cost Forwarding Algorithm

The Minimum Cost Forwarding Algorithm(MCFA) algorithm exploits the fact that the direction of routing is always known, that is, toward the fixed external BS. Hence, a sensor node need not have a unique ID nor maintain a routing table. Instead, each node maintains the least cost estimate from itself to the BS. Each message to be forwarded by the sensor node is broadcast to its neighbors. When a node receives the message, it checks if it is on the least cost path between the source sensor node and the BS. If this is the case, it rebroadcasts the message to its neighbors. This process repeats until the BS is reached.[29]

In MCFA, each node should know the least cost path estimate from itself to the BS. This is obtained as follows. The sink broadcasts a message with the cost set to zero, while every node initially set its least cost to the sink to infinity (∞). Each node, upon receiving the broadcast message originated at the sink, checks to see if the estimate in the message andthe link on which it is received is less than the current estimate. If yes, the current estimate and the estimate in the broadcast message are updated. If the received broadcast message is updated, then it is resent, otherwise, it is purged and nothing further is done.[29]

However, the previous procedure may result in some nodes having multiple updates and those nodes far away from the sink will get more updates from those closer to the sink. To avoid this, the MCFA was modified to run a back-off algorithm at the setup phase. The back-off algorithm dictates that a node will not send the updated message until a*lctime units have elapsed from the time at which the message is updated, where a is a constant and lc is the link cost from which the message was received.[29]

4.3.1.2   HIERARCHICAL ROUTING

Hierarchical protocols aim at clustering the nodes so that cluster heads (CHs) can do some data aggregation and reduction in order to save energy (see Fig. 4.8). Hierarchical routing is mainly two-layer routing where one layer is used to select CHs, and another for routing. Nodes in hierarchical networks play different roles. Higher-energy nodes can be used to process and send information; low-energy nodes can be used to perform the sensing in the proximity of the target. Hierarchical routing is utilized to perform energy-efficient routing in WSNs. Hierarchical routing is an efficient method for lower energy consumption within a cluster.[11]

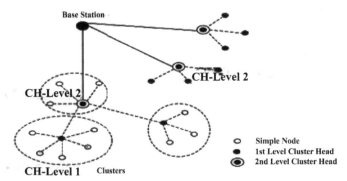

FIGURE 4.8   Cluster-Based Hierarchical Model.
Source: Ref [15] with permission.

4.3.1.2.1   Low-Energy Adaptive Clustering Hierarchy

Low-Energy Adaptive Clustering Hierarchy (LEACH) is the first and most popular energy-efficient hierarchical clustering algorithm for WSNs that was proposed for reducing power consumption. In LEACH, the clustering task is rotated amongthe nodes, based on duration. Direct communication is used by each CH to forward the data to the sink. It uses clusters to prolong the life of the WSN. Optimal number of CHs is estimated to be 5% of the total number of nodes (due to Heinzelman, Chandrakasan, and Balakrishnan simulation model).

LEACH is based on an aggregation technique that combines or aggregates the original data into a smaller size of data that carry only meaningful information to all individual sensors. LEACH divides a network into several clusters of sensors, which are constructed by using localized coordination and control not only to reduce the amount of data that are transmitted to the sink but also to make routing and data dissemination more scalable and robust (see Fig. 4.9). LEACH uses a randomize rotation of high-energy CH position rather than selecting in static manner, to give a chance to all sensors to act as CHs and avoid the battery depletion of an individual sensor and dying quickly. The operation of LEACH is divided into rounds having two phases; the first is a setup phase to organize the network into clusters, CH advertisement, and transmission schedule creation and the second is a steady-state phase for data aggregation, compression, and transmission to the sink.[25]

LEACH is completely distributed and requires no global knowledge of network. It reduces energy consumption by minimizing the communication cost between sensors and their CHs and turning off nonhead nodes as much as possible.

LEACH uses single-hop routing where each node can transmit directly to the CH and the sink. Therefore, it is not applicable to networks deployed in large regions. Furthermore, the idea of dynamic clustering brings extra overhead, for example, head changes, advertisements, and so forth, which may diminish the gain in energy consumption. While LEACH helps the sensors within their cluster dissipate their energy slowly, the CHs consume a larger amount of energy when they are located farther away from the sink. Also, LEACH clustering terminates in a finite number of iterations, but does not guarantee good CH distribution and assumes uniform energy consumption for CHs.[25]

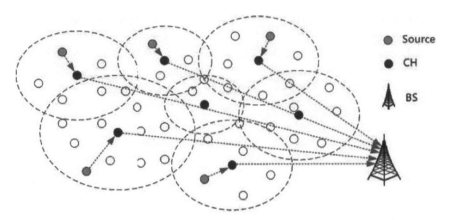

FIGURE 4.9 Low-Energy Adaptive Clustering Hierarchy communication architecture.
Source: Ref [15] with permission.

4.3.1.2.2   *Power-Efficient Gathering in Sensor Information Systems*

The basic idea of this protocol is that in order to extend network lifetime, nodes need only communicate with their closest neighbors and they take turns in communicating with the sink. When the round of all nodes communicating with the sink ends, a new round will start and so on. This reduces the power required to transmit data per round as the power draining is spread uniformly over all nodes. Hence, Power-Efficient Gathering in Sensor Information Systems(PEGASIS) has two main objectives. First, increase the lifetime of each node by using collaborative techniques and as a result the network lifetime will be increased. Second, allow only local coordination between nodes that are close together so that the bandwidth consumed in communication is reduced.

Unlike LEACH, PEGASIS avoids cluster formation and uses only one node in a chain to transmit to the sink instead of using multiple nodes.[20] In Figure 4.10, node c2 is the leader, and it will pass the token along the chain to node c0. Node c0 passes its data to node c1. Node c1 aggregates node c0's data with its own and then transmits to the leader. After node c2 passes the token to node c4, node c4 transmits its data to node c3. Node c3 aggregates node c4's data with its own and then transmits to the leader. Node c2 waits to receive data from both neighbors and then aggregates its data with its neighbors' data. Finally, node c2 transmits one message to the sink.

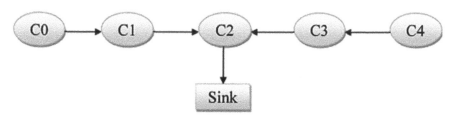

FIGURE 4.10   Chaining in Power-Efficient Gathering in Sensor Information Systems. *Source:* Ref [15] with permission.

To locate the closest neighbor node in PEGASIS, each node uses the signal strength to measure the distance to all neighboring nodes and then adjusts the signal strength so that only one node can be heard. The chain in PEGASIS will consist of those nodes that are closest to each other and form a path to the BS. The aggregated form of the data will be sent to the BS by any node in the chain and the nodes in the chain will take turns in sending to the BS. The chain construction is performed in a greedy fashion.[3]

PEGASIS has been shown to outperform LEACH by about 100–300% for different network sizes and topologies. Such performance gain is achieved through the elimination of the overhead caused by dynamic cluster formation in LEACH and through decreasing the number of transmissions and reception by using data aggregation.[1]

However, PEGASIS still requires dynamic topology adjustment since a sensor node needs to know about energy status of its neighbors in order to know where to route its data. Such topology adjustment can introduce significant overhead especially for highly utilized networks. Moreover, PEGASIS assumes that each sensor node isable to communicate with the BS directly. In practical cases, sensor nodes use multi-hop communication to reach the BS. Note also that PEGASIS introduces excessive delay for distant node on the chain. In addition, the single leader can become a bottleneck.[3]

An extension to PEGASIS, called Hierarchical-PEGASIS was introduced with the objective of decreasing the delay incurred for packets during transmission to the sink. For this purpose, simultaneous transmissions of data are studied in order to avoid collisions through approaches that incorporate signal coding and spatial transmissions.[3]

4.3.1.2.3   Threshold-sensitive Energy-Efficient Sensor Network and Adaptive Periodic Threshold-sensitive Energy-Efficient Sensor Network

Threshold-sensitive Energy-Efficient sensor Network (TEEN) and Adaptive Periodic Threshold-sensitive Energy-Efficient sensor Network (APTEEN) were proposed for time-critical applications, in which the network operated in a reactive mode. In TEEN, sensor nodes sense the medium continuously, but the data transmission is done less frequently. A CH sensor sends its members a hard threshold, which is the threshold value of the sensed attribute and a soft threshold, which is a small change in the value of the sensed attribute that triggers the node to switch on its transmitter and transmit.

Thus, the hard threshold tries to reduce the number of transmissions by allowing the nodes to transmit only when the sensed attribute is in the range of interest.[3] The soft threshold further reduces the number of transmissions that might have otherwise occurred when there is little or no change in the sensed attribute a smaller value of the soft threshold gives a more accurate picture of the network, at the expense of increased energy consumption. Thus, the user can control the trade-off between energy efficiency and data accuracy.[3] When CHs are to change (Fig. 4.11), new values for the above parameters are broadcast. The main drawback of this scheme is that if the thresholds are not received, the nodes will never communicate, and the user will not get any data from the network at all.

The nodes sense their environment continuously. The first time a parameter from the attribute set reaches its hard threshold value, the node switches its transmitter on and sends the sensed data. The sensed value (SV) is stored in an internal variable, called SV. The nodes will transmit data in the current cluster period only when the following conditions are true: (1) The current value of the sensed attribute is greater than the hard threshold. (2) The current value of the sensed attribute differs from SV by an amount equal to or greater than the soft threshold.[3]

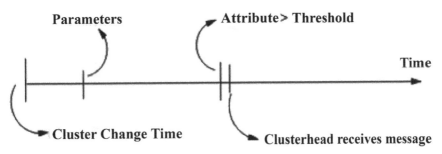

FIGURE 4.11 Operation of Threshold-sensitive Energy-Efficient Sensor Network. *Source:* Ref [15] with permission.

The important features of TEEN include its suitability for time-critical sensing applications. Moreover, message transmission consumes more energy than data sensing. Therefore, even though the nodes sense continuously, the energy consumption in this scheme can potentially be much less than in the proactive network, because data transmission is done less frequently. The soft threshold can be varied. At every cluster change time, fresh parameters are broadcast and therefore the user can change them as required.[21]

APTEEN, on the other hand, is an improvement to TEEN to overcome its shortcomings, and it is a hybrid protocol that aims at both capturing periodic data collections (LEACH) and reacting to time-critical events (TEEN). APTEEN supports three different query types: (1) historical query, to analyze past data values;(2) one-time query, to take a snapshot view of the network; and (3) persistent queries, to monitor an event for a period of time.[1]

In APTEEN, the CHs broadcast the attributes, the threshold values, the transmission schedule, and the count time (i.e., the maximum time period between two successive reports sent by a node). The node senses the environment continuously, and nodes transmit data, only when the previous two conditions are verified. If a node does not send data for a time period equal to the count time (see Fig. 4.12), it is forced to sense and retransmit the data.[3]

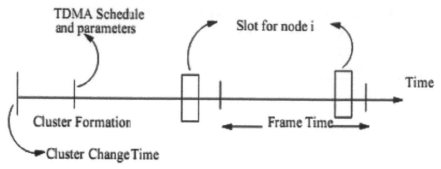

FIGURE 4.12 Operation of Adaptive Periodic Threshold-sensitive Energy-Efficient sensor Network.
Source: Ref [15] with permission.

The main features of the APTEEN scheme include the combination of both proactive and reactive policies. It offers a lot of flexibility by allowing the user to set the count time interval, and the threshold values for the energy consumption can be controlled by changing the count time as well as the threshold values. The main drawback of the scheme is the additional complexity required to implement the threshold functions and the count time.[3]

The simulation of TEEN and APTEEN has shown that these two protocols outperform LEACH. The experiments have demonstrated that APTEEN's performance is somewhere between LEACH and TEEN in terms of energy dissipation and network lifetime. TEEN gives the best performance since it decreases the number of transmissions. The main drawbacks of the two approaches are the overhead and complexity associated with forming clusters at multiple levels, the method of implementing threshold-based functions, and how to deal with attribute-based naming of queries.[3]

4.3.1.2.4 Hybrid Energy-Efficient Distributed Clustering

Hybrid Energy-Efficient Distributed Clustering (HEED) extends the basic scheme of LEACH by using residual energy and node degree or density as a metric for cluster selection to achieve power balancing. It operates in multi-hop networks, using an adaptive transmission power in the inter-clustering communication. HEED was proposed with four primary goals, namely, (1) prolonging network lifetime by distributing energy consumption, (2) terminating the clustering process within a constant number of iterations, (3) minimizing control overhead, and (4) producing well-distributed CHs and compact clusters.[30]

In HEED, the proposed algorithm periodically selects CHs according to a combination of two clustering parameters. The primary parameter is their residual energy of each sensor node (used in calculating probability of becoming a CH) and the secondary parameter is the intra-cluster communication cost as a function of cluster density or node degree (i.e., number of neighbors).

The primary parameter is used to probabilistically select an initial set of CHs, while the secondary parameter is used for breaking ties.[30]

The HEED clustering improves network lifetime over LEACH clustering because LEACH randomly selects CHs (and hence cluster size), which may result in faster death of some nodes. The final CHs selected in HEED are well distributed across the network and the communication cost is minimized. However, the cluster selection deals with only a subset of parameters, which can possibly impose constraints on the system. These methods are suitable for prolonging the network lifetime rather than for the entire needs of WSNs.[30]

4.3.1.2.5 Self-Organizing Protocol

Subramanian and Katz describe a self-organizing protocol and an application taxonomy that was used to build architecture used to support Heterogeneous sensors. Furthermore, these sensors can be mobile or stationary. Some sensors probe the environment and forward the data to a designated set of nodes that act as routers. Router nodes are stationary and form the backbone for communication. Collected data are forwarded through the routers to more powerful sink nodes. Each sensing node should be reachable to a router node in order to be part of the network.[1]

A routing architecture that requires addressing of each sensor node has been proposed. Sensing nodes are identifiable through the address of the router node it is connected to. The routing architecture is hierarchical where groups of nodes are formed and merge when needed. In order to support fault tolerance, local Markov loops (LML) algorithm, which performs a random walk on spanning trees of a graph, is used in broadcasting. The algorithm for self-organizing the router nodes and creating the routing tables consists of four phases:[27]

- **Discovery phase:** The nodes in the neighborhood of each sensor are discovered.
- **Organization phase:** Groups are formed and merged by forming a hierarchy. Each node is allocated an address based on its position in the hierarchy. Routing tables are created for each node. Broadcast trees that span all the nodes are constructed.
- **Maintenance phase:** Updating of routing tables and energy levels of nodes is made in this phase. Each node informs the neighbors about its routing table and energy level. LML are used to maintain broadcast trees.
- **Self-reorganization phase:** In case of partition or node failures, group reorganizations are performed.

In this approach, sensor nodes can be addressed individually in the routing architecture, and hence it is suitable for the applications where communication to a particular node is required. Furthermore, this algorithm incurs a small cost for maintaining routing tables and keeping a balanced routing hierarchy.

It was also found that the energy consumed for broadcasting a message is less than that consumed in the SPIN protocol. This protocol, however, is not an on-demand protocol especially in the organization phase of algorithm. Another issue is related to the formation of hierarchy. It could happen that there are many cuts in the network, and hence the probability of applying reorganization phase increases, which will be an expensive operation.[3]

4.3.1.2.6   Hierarchical Power-Aware Routing

The protocol divides the network into groups of sensors. Each group of sensors in geographic proximity is clustered together as a zone, and each zone is treated as an entity. To perform routing, each zone is allowed to decide how it will route a message hierarchically across the other zones such that the battery lives of the nodes in the system are maximized.

Messages are routed along the path which has the maximum over all the minimum of the remaining power, called the max–min path. The motivation is that using nodes with high residual power may be expensive as compared withthe path with the minimal power consumption.[19] An approximation algorithm, called the max–min zPmin algorithm, was proposed. The crux of the algorithm is based on the trade-off between minimizing the total power consumption and maximizing the minimal residual power of the network. Hence, the algorithm tries to enhance a max–min path by limiting its power consumption as follows. First, the algorithm finds the path with the least power consumption (Pmin) by using the Dijkstra algorithm. Second, the algorithm finds a path that maximizes the minimal residual power in the network. The proposed algorithm tries to optimize both solution criteria. This is achieved by relaxing the minimal power consumption for the message to be equal to zPmin with parameter $z \geq 1$ to restrict the power consumption for sending one message to zPmin. The algorithm consumes at most zPmin while maximizing the minimal residual power fraction.[19]

4.3.1.3   DIRECT ROUTING

In this taxonomy of protocols, all the sensor nodes communicate directly with the sink. This requires a renewable energy source because of the important energy consumption due to the direct communications. This type is rarely used in sensor networks because of its dense nature where the nodes are randomly deployed and equipped with a small battery. Furthermore, the direct communication considerably reduces the network lifetime.[2]

4.3.2   ROUTING PROTOCOLS BASED ON PROTOCOL OPERATION

In this section, we will review routing protocols that have different routing functionalities.

4.3.2.1   MULTIPATH ROUTING PROTOCOLS

In this subsection, we study the routing protocols that use multiple paths rather than a single path in order to enhance the network performance. The fault tolerance (resilience) of a protocol is measured by the likelihood that an alternate path exists between a source and a destination when the primary path fails. This can be increased by maintaining multiple paths between the source and the destination at the expense of an increased energy consumption and traffic generation. These alternate paths are kept alive by sending periodic messages. Hence, network reliability can be increased at the expense of increased overhead of maintaining the alternate paths.[3]

In this context, a lot of algorithms are proposed:

- Chang and Tassiulas proposed an algorithm which will route data through a path whose nodes have the largest residual energy. The path is changed whenever a better path is discovered. The primary path will be used until its energy falls below the energy of the backup path at which the backup path is used. Using this approach, the nodes in the primary path will not deplete their energy resources through continual use of the same route, hence achieving longer life.[46]

- Rahul and Rabaey proposed the use of a set of suboptimal paths occasionally to increase the lifetime of the network. These paths are chosen by means of a probability which depends on how low the energy consumption of each path is; the path with the largest residual energy when used to route data in a network may be very energyexpensive. Therefore, there is a trade-off between minimizing the total power consumed and the residual energy of the network.[3]

- Li, Aslam, and Rus proposed an algorithm in which the residual energy of the route is relaxed a bit in order to select a more energy-efficient path.[3]

Directed Diffusion is a good candidate for robust multipath routing and delivery. It has been found that the use of multipath routing provides viable alternative for energy-efficient recovery from failures in WSNs. The motivation of using these braided paths is to keep the cost of maintaining the multipaths low.

The costs of alternate paths are comparable to the primary path because they tend to be much closer to the primary path.

4.3.2.2   QUERY-BASED ROUTING

In this kind of routing, the destination nodes propagate a query for data (sensing task or interest) from a node through the network and a node having this data sends the data which matches the query back to the node, which initiates the query. Usually, these queries are described in natural language or in high-level query languages. For example, client C1 may submit a query to node N1 and ask: Are there moving vehicles in battle space region1? All the nodes have tables consisting of the sensing tasks queries that they receive and send data which matches these tasks when they receive it.[3]

In Directed Diffusion, the sink node sends out interest messages to sensors. As the interest is propagated throughout the sensor network, the gradients from the source back to the sink are setup. When the source has data for the interest, the source sends the data along the interest's gradient path. To lower energy consumption, data aggregation (e.g., duplicate suppression) is performed en route.

The rumor routing protocol uses a set of long-lived agents to create paths that are directed toward the events they encounter. Whenever an agent crosses path with a path leading to an event that it has not encountered yet, it creates a path state that leads to the event. When the agents come across shorter paths or more efficient paths, they optimize the paths in routing tables accordingly. Each node maintains a list of its neighbors and an events table that is updated whenever new events are encountered. Each node can also generate an agent in a probabilistic fashion. Each agent contains an events table that is synchronized with every node that it visits. The agent has a lifetime of a certain number of hops after which it dies. A node will not generate a query unless it learns a route to the required event. If there is no route available, the node transmits a query in a random direction. Then, node waits to know if the query reached the destination for a certain amount of time, after which the node floods the network if no response is heard from the destination.

4.3.2.3   NEGOTIATION-BASED ROUTING PROTOCOLS

These protocols use high-level data descriptors in order to eliminate redundant data transmissions through negotiation. Communication decisions are also taken based on the resources that are available to them.[3]

The SPIN family protocols are examples of negotiation-based routing protocols. The motivation is that the use of flooding to disseminate data will produce implosion and overlap between the sent data; hence, nodes will receive duplicate copies of the same data.

This operation consumes more energy and more processing by sending the same data by different sensors. The SPIN protocols are designed to disseminate the data of one sensor to all other sensors assuming these sensors are potential BSs. Hence, the main idea of negotiation-based routing in WSNs is to suppress duplicate information and prevent redundant data from being sent to the next sensor or the sink by conducting a series of negotiation messages before the real data transmission begins.

4.3.2.4   QOS-BASED ROUTING

In QoS-based routing protocols, the network has to balance between energy consumption and data quality. In particular, the network has to satisfy certain QoS metrics, for example, delay, energy, bandwidth, and so forth, when delivering data to the sink.[3]

4.3.2.4.1   Sequential Assignment Routing

Sequential Assignment Routing (SAR) is the first protocol for sensor networks that includes the notion of QoS in its routing decisions. It is a table-driven multipath approach striving to achieve energy efficiency and fault tolerance. The SAR protocol creates trees rooted at one-hop neighbors of the sink by taking QoS metric, energy resource on each path, and priority level of each packet into consideration.[16] By using created trees, multiple paths from sink to sensors are formed. One of these paths is selected according to the energy resources and QoS on the path. Failure recovery is done by enforcing routing table consistency between upstream and downstream nodes on each path. Any local failure causes an automatic path restoration procedure locally. Simulation results show that SAR offers less power consumption than the minimum-energy metric algorithm, which focuses only the energy consumption of each packet without considering its priority. SAR maintains multiple paths from nodes to sink. Although this ensures fault tolerance and easy recovery, the protocol suffers from the overhead of maintaining the tables and states at each sensor node especially when the number of nodes is huge.[26]

4.3.2.4.2   SPEED

A Stateless Protocol for Real-Time Communication in Sensor Networks (SPEED), a real-time communication protocol, is another QoS routing protocol for sensor networks that provides soft real-time end-to-end guarantees. The protocol requires each node to maintain information about its neighbors and uses geographic forwarding to find the paths. In addition, SPEED strives to ensure a certain speed for each packet in the network so that each application can estimate the EED for the packets by dividing the distance to the sink by the speed of the packet before making the admission decision.

Moreover, SPEED can provide congestion avoidance when the network is congested, and the total transmission energy is less due to the simplicity of the routing algorithm, that is, control packet overhead is less, and to even traffic distribution.[13]

The routing module in SPEED is called stateless geographic nondeterministic forwarding (SNGF) and works with four other modules at the network layer, as shown in Figure 4.13.

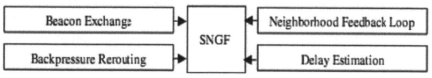

FIGURE 4.13   Routing components of SPEED.
Source: Ref [15] with permission.

The beacon exchange mechanism collects information about the nodes and their location. Delay estimation at each node is basically made by calculating the elapsed time when an acknowledgment is received from a neighbor as a response to a transmitted data packet. By looking at the delay values, SNGF selects the node, which meets the speed requirement. If such a node cannot be found, the relay ratio of the node is checked. The neighborhood feedback loop module is responsible for providing the relay ratio which is fed to the SNGF module. If the relay ratio is less than a randomly generated number between 0 and 1, the packet is dropped. And finally, the backpressurererouting module is used to prevent voids when a node fails to find a next hop node, and to eliminate congestion by sending messages back to the source nodes so that they will pursue new routes.[1]

4.3.2.4.3   MultiPath and Multi-SPEED

An extension of SPEED is designed with two important goals:(1) localized packet routing decision without global network state update or a priori path setup and (2) providing differentiated QoS options in timeliness and reliability domains. For the localized packet routing without end-to-end path setup and maintenance, we adopt the geographic routing mechanism based on location awareness. First, we assume that the packet destination is specified by a geographic location rather than node ID. Also, each sensor node is assumed to be aware of its geographical location using GPS. The location information can be exchanged with immediate neighbors with "periodic location update packets." Thus, each node is aware of its immediate neighbors within its radio range and their locations. Using the neighbor locations, each node can locally make a per-packet routing decision such that packets progress geographically toward their final destinations. If each node relays the packet to a neighbor closer to the destination area, the packet can eventually be delivered to the destination without global topology information.[12]

4.3.2.5   COHERENT AND NONCOHERENT PROCESSING

Data processing is a major component in the operation of WSNs. Hence, routing techniques employ different data processing techniques. In general, sensor nodes will cooperate with each other in processing different data flooded in the network area. Two examples of data processing techniques proposed in WSNs are coherent and noncoherent data-processing-based routing.[3]

In noncoherent data processing routing, nodes will locally process the raw data before being sent to other nodes for further processing. The nodes that perform further processing are called the aggregators. In noncoherent processing, data processing incurs three phases:[3]

- Target detection, data collection, and preprocessing.
- Membership declaration: When a node decides to participate in a cooperative function, it declares its intention to all neighbors. This should be done as soon as possible so that each sensor has a local understanding of the network topology.
- Central node election: It must have sufficient energy reserves and computational capability because it is selected to perform more sophisticated information processing.

In coherent routing, the data is forwarded to aggregators after minimum processing. The minimum processing typically includes tasks such astime stamping, duplicate suppression, and so forth. To perform energy-efficient routing, coherent processing is normally selected.[3]

4.3.3 COMMUNICATION PARADIGM-BASED PROTOCOLS

The communication paradigm determines how the nodes are interrogated. In WSNs, we can distinguish three communication paradigms.

4.3.3.1 NODE CENTRIC

It is used in conventional networks where it is necessary to know and identify communicating nodes. Ad hoc networks use this paradigm which fits such type of environment. However, in WSNs, routing based on individual nodes identification does not reflect the real use of WSNs. However,node-centric paradigm is still used in some WSNs applications requiring individual node interrogation.

4.3.3.2   DATA CENTRIC

In many applications of sensor networks, it is not feasible to assign global identifiers to each node due to the sheer number of nodes deployed. Such lack of global identification along with random deployment of sensor nodes makes it hard to select a specific set of sensor nodes to be queried. Therefore, data is usually transmitted from every sensor node within the deployment region with significant redundancy. Since this is very inefficient in terms of energy consumption, routing protocols that will be able to select a set of sensor nodes and utilize data aggregation during the relaying of data have been considered.

This consideration has led to data-centric routing, which is different from traditional address-based routing, where routes are created between addressable nodes managed in the network layer of the communication stack.

SPIN is the first data-centric protocol, which considers data negotiation between nodes in order to eliminate redundant data and save energy. Later, Directed Diffusion has been developed and has become a breakthrough in data-centric routing. Then, many other protocols have been proposed either based on Directed Diffusion or following a similar concept.[1]

4.3.3.3   POSITION CENTRIC

Most of the routing protocols for sensor networks require location information for sensor nodes. In most cases, location information is needed in order to calculate the distance between two particular nodes so that energy consumption can be estimated. Since there is no addressing scheme for sensor networks like IPaddresses and they are spatially deployed on a region, location information can be utilized in routing data in an energy-efficient way. For instance, if the region to be sensed is known, using the location of sensors, the query can be diffused only to that particular region which will eliminate the number of transmission significantly.

Some of the protocols discussed here are designed primarily for MANET, considering the mobility of nodes during the design. However, they are also well applicable to sensor networks where there is less or no mobility.

4.3.3.3.1 Minimum-Energy Communication Network and Minimum-Energy Communication Network

Minimum-Energy Communication Network (MECN) sets up and maintains a minimum-energy network for wireless networks by utilizing low-power GPS. The main idea of MECN is to find a subnetwork, which will have less number of nodes and require less power for transmission between any two particular nodes.

In this way, global minimum power paths are found without considering all the nodes in the network. MECN is self-reconfiguring and thus can dynamically adapt to nodes failure or the deployment of new sensors.[1]

The Small Minimum-Energy Communication Network (SMECN) is an extension to MECN. In MECN, it is assumed that every node can transmit to every other node, which is not possible every time. In SMECN, possible obstacles between any pair of nodes are considered. However, the network is still assumed to be fully connected as in the case of MECN. The subnetwork constructed by SMECN for minimum-energy relaying is provably smaller (in terms of number of edges) than the one constructed in MECN if broadcasts are able to reach to all nodes in a circular region around the broadcaster. As a result, the number of hop for transmissions will decrease.[1]

Simulation results show that SMECN uses less energy than MECN and maintenance cost of the links is less. However, finding a subnetwork with smaller number of edges introduces more overhead in the algorithm.[22]

4.3.3.3.2 Geographic Adaptive Fidelity

Geographic adaptive fidelity (GAF) is an energy-aware location-based routing algorithm designed primarily for MANET, but may be applicable to sensor networks as well. GAF conserves energy by turning off unnecessary nodes in the network without affecting the level of routing fidelity. It forms a virtual grid for the covered area. Each node uses its GPS-indicated location to associate itself with a point in the virtual grid.[3]

Nodes associated with the same point on the grid are considered equivalent in terms of the cost of packet routing. Such equivalence is exploited in keeping some nodes located in a particular grid area in sleeping state in order to save energy. Thus, GAF can substantially increase the network lifetime as the number of nodes increases.

A sample situation is depicted in Figure 4.14. In this figure, node 1 can reach any of 2, 3, and 4 and nodes 2, 3, and 4 can reach 5. Therefore, nodes 2, 3, and 4 are equivalent and two of them can sleep.[28]

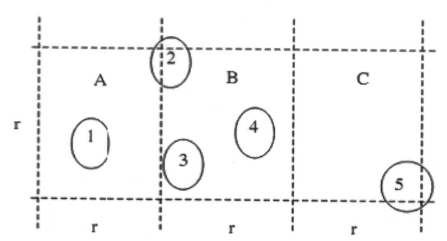

FIGURE 4.14 Example of a virtual grid in geographic adaptive fidelity (GAF). *Source:* Ref [15] with permission.

Nodes change states from sleeping to active in turn so that the load is balanced. There are three states defined in GAF. These states are discovery, for determining the neighbors in the grid, active reflecting participation in routing, and sleep when the radio is turned off. The state transitions in GAF are depicted in Figure 4.15. Which node will sleep for how long is application dependent and the related parameters are tuned accordingly during the routing process.[28]

In order to handle the mobility, each node in the grid estimates its leaving time of grid and sends this to its neighbors. The sleeping neighbors adjust their sleeping time accordingly in order to keep the routing fidelity. Before the leaving time of the active node expires, sleeping nodes wake up and one of them becomes active. GAF is implemented both for nonmobility (GAF-basic) and mobility (GAF-mobility adaptation) of nodes.[1]

Simulation results show that GAF performs at least as well as a normal ad hoc routing protocol in terms of latency and packet loss and increases the lifetime of the network by saving energy.[28]

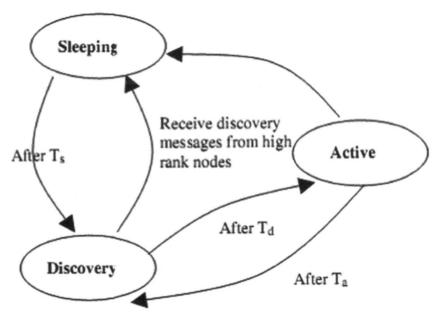

FIGURE 4.15   State transitions in GAF.
Source: Ref [15] with permission.

4.3.3.3.3   *Geographic and Energy-Aware Routing*

Geographic and Energy-Aware Routing (GEAR) uses energy-aware and geographically informed neighbor selection heuristics to route a packet toward the target region. The idea is to restrict the number of interests in Directed Diffusion by only considering a certain region rather than sending the interests to the whole network. GEAR compliments Directed Diffusion in this way and thus conserves more energy.[3]

   In GEAR, each node keeps an estimated cost and a learning cost of reaching the destination through its neighbors. The estimated cost is a combination of residual energy and distance to destination. The learned cost is a refinement of the estimated cost that accounts for routing around holes in the network. A hole occurs when a node does not have any closer neighbor to the target region than itself. If there are no holes, the estimated cost is equal to the learned cost. The learned cost is propagated one hop back every time a packet reaches the destination so that route setup for next packet will be adjusted. There are two phases in the algorithm:[31]

- **Forwarding packets toward the** target region: Upon receiving a packet, a node checks its neighbors to see if there is one neighbor, which is closer to the target region than itself. If there is more than one, the nearest neighbor to the target region is selected as the next hop. If they are all further than the node itself, this means there is a hole. In this case, one of the neighbors is picked to forward the packet based on the learning cost function.
- **Forwarding the packets within the region:**If the packet has reached the region, it can be diffused in that region by either recursive geographic forwarding or restricted flooding. Restricted flooding is good when the sensors are not densely deployed. In high-density networks, recursive geographic flooding is more energy efficient than restricted flooding.

In that case, the region is divided into four subregions and four copies of the packet are created. This splitting and forwarding process continues until the regions with only one node are left. An example is depicted in Figure 4.16.

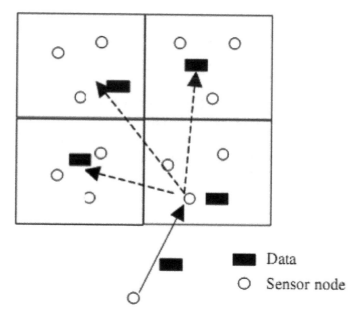

FIGURE 4.16 Recursive geographic forwarding in Geographic and Energy-Aware Routing.

Source: Ref [15] with permission.

4.3.4   ROUTES ESTABLISHMENT-BASED PROTOCOLS

Depending on route creation and maintaining, routing protocols can be classified into three categories: reactive, proactive, and hybrid protocols.

4.3.4.1   REACTIVE ROUTING PROTOCOLS

In reactive routing protocol, routing actions are triggered when there is data to be sent and disseminated to other nodes. Here, paths are setup on demand when queries are initiated. A route search is needed for every new destination; therefore, the communication overhead is reduced at the expense of delay to search the route.[9]

4.3.4.1.1   Dynamic Source Routing

To send a packet to another sensor, the sender constructs a source route in the packet's header, giving the address of each node in the network through which the packet should be forwarded in order to reach the destination node. The sender then transmits the packet over its wireless network interface to the first hop identified in the source route. When a node receives a packet, if this one is not the final destination of the packet, it simply transmits the packet to the next hop identified in the source route in the packet's header until it reaches the destination.[17]

Dynamic Source Routinghas two main operations]:[17]

Route Discovery:Route discovery allows any node in the network to dynamically discover a route to any other one, whether directly reachable within wireless transmission range or reachable through one or more intermediate network hops through other nodes. A node initiating a route discovery broadcasts a route request packet (Fig. 4.17), which may be received by those nodes within wireless transmission range of it. The route request packet identifies the destination node, referred to as the target of the route discovery, for which the route is requested. If the route discovery is successful, the initiating host receives a route reply packet (Fig. 4.18) listing a sequence of network hops through which it may reach the target.

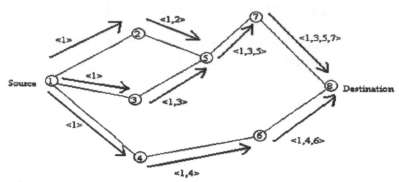

FIGURE 4.17   Route request packet broadcast.
Source: Ref [15] with permission.

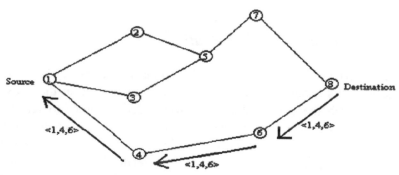

FIGURE 4.18   Reply packet.
Source: Ref [15] with permission.

Route Maintenance: To ensure the data and the route arein use validly, the route maintenance procedure monitors the operation of this route and informs the sender of any routing errors.

Each intermediate node has the possibility to maintain a route cache in which it caches source routes that it has learned. When the sensor wants to send a packet; it first checks its route cache for a source route to the destination. If a route is found, the sender uses this route to transmit the packet. If no route is found, the sender may attempt to discover one using the route discovery operation. If any of the intermediate nodes along the route fail or arepowered off, the route can no longer be used to reach the destination. Therefore, while a node is using any source route, it monitors the continued correct operation of that route using the route maintenance operation.

4.3.4.2   HYBRID ROUTING PROTOCOLS

There exist a number of routing protocols of globally reactive and locally proactive states. The routing table of each node contains information about only the nearest neighbors. Therefore, during route research, these protocols consult the table routing for the nearest neighbors and use reactive methods beyond.[18]

4.3.4.2.1   Zone Routing Protocol

Zone routing protocol (ZRP) can be classed as a hybrid reactive/proactive routing protocol. ZRP aims to address the problems by combining the best properties of both the approaches. The ZRP, as its name implies, is based on the concept of zones. A routing zone is defined for each node separately, and the zones of neighboring nodes overlap. The routing zone has a radius ρ expressed in hops. The zone thus includes the nodes, whose distance from the node in question is at most ρ hops.

An example routing zone is shown in Figure 4.19, where the routing zone of S includes the nodes A–I, but not K. In the illustrations, the radius is marked as a circle around the node in question. It should, however, be noted that the zone is defined in hops, not as a physical distance.[5]

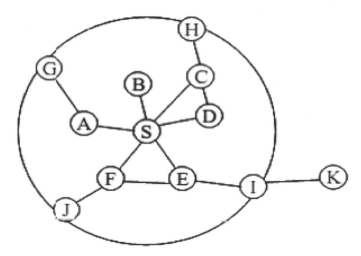

FIGURE 4.19   Example of routing zone with ρ=2.
Source: Ref [15] with permission.

The nodes of a zone are divided into peripheral nodes and interior nodes. Peripheral nodes are nodes whose minimum distance to the central node is exactly equal to the zone radius ρ.

The nodes whose minimum distance is less than ρ are interior nodes. In Figure 4.19, the nodes A–F are interior nodes; the nodes G–J are peripheral nodes.

4.4   CONCLUSION

Routing in sensor networks has attracted a lot of attention in the recent years and introduced unique challenges compared to traditional data routing in wired networks.

Several routing protocols have been proposed for WSNs.[1,4] Each protocol is adapted to a specific situation and must take into account the type of application. In this chapter, we have tried to summarize recent research results on data routing in sensor networks and classified them according to the approaches they use. Our classification is based on four main categories, namely network structure, protocol operating, communication paradigm, and routes establishment.

KEYWORDS

- **mobile ad hoc networks**
- **available energy**
- **required energy**
- **end-to-end delay**
- **Sensor Protocol for Information via Negotiation**

REFERENCES

1. Akkaya, K.; Younis, M. A. Survey on Routing Protocols for Wireless Sensor Networks. *Ad-Hoc Netw. J.***2005**,*3*(3) 326–349.
2. Allam, S. *Approche Multi agents pour contrôler l'inondation dans un réseau de capteurs,* Memory to obtain the engineering degree in informatics; École nationale Supérieure d'Informatique, 2009.

3. Al-karaki, J. N.; Kamal, A. E. *Routing Techniques in Wireless Sensor Networks: A Survey*; Dept. of Electrical and Computer Engineering Iowa State University: Iowa, USA, 2003.
4. Al-Karaki J.N.; Kamal, A.E. Routing Techniques in Wireless Sensor Networks: A Survey. *IEEE Wirel. Commun.* **2004**,*6*(11), 6–28.
5. Beijar, N. Zone Routing Protocol (ZRP). Networking Laboratory, Helsinki University of Technology, Finland.
6. Berrachedi, A.; Diarbakirli, A. *Sécurisation du protocole de routage hiérarchique LEACH dans les réseaux de capteurs sans fil, Memory to obtain the engineering degree in informatics*; École nationale Supérieure d'Informatique, 2009.
7. Braginsky, D.; Estrin, D. In *Rumor Routing Algorithm for Sensor Networks,* Proceedings of the First Workshop on Sensor Networks and Applications (WSNA), Atlanta, GA, USA, October 2002.
8. Chang , J. H.; Tassiulas, L. In *Maximum Lifetime Routing in Wireless Sensor Networks*, Proceedings Advanced Telecommunications and Information Distribution Research Program (ATIRP2000), College Park, MD, USA, March 2000.
9. Chen, J. L.; Hsu, Y. M.; Chang, I. C. Adaptive Routing Protocol for Reliable Sensor Network Applications. *Int. J. Smart Sens. Intell. Syst.***2009**,*2*(4), 515-539.
10. Chu, M.; Haussecker, H.; Zhao, F. Scalable Information-Driven Sensor Querying and Routing for ad hoc Heterogeneous Sensor Networks. *Int. J. High Perform. Comput. Appl.* **2002**, *16*(3), 293-313.
11. Dwivedi, A. K.; Vyas, O. P. Network Layer Protocols for Wireless Sensor Networks: Existing Classifications and Design Challenges. *Int. J. Comput. Appl.***2010**,*8*(12), pp.30–34.
12. Felemban, E.; Chang-gun, L.; Ekici, E. MMSPEED: Multipath Multi-SPEED Protocol for QoS Guarantee of Reliability and Timeliness in Wireless Sensor Networks. IEEE Trans. *Mobile Comput.***2006**,*5*, 738–754.
13. He, T.; Stankovic, J. A.; Abdelzaher, T. F.; Lu, C. In *A Spatiotemporal Communication Protocol for Wireless Sensor Networks,* Proceedings of International Conference on Distributed Computing Systems, Providence, Rhode Island, USA, May 2003.
14. Heinzelman, W.; Kulik, J.; Balakrishnan, H. *Adaptive Protocols for Information Dissemination in Wireless Sensor Networks*; Research report, Massachusetts Institute of Technology Cambridge: USA, 1999.
15. Hibi, A. WSNs Routing Protocols. (Master Thesis 2011).
16. Houngbadji, T. *Réseaux Ad-Hoc: Système d'adressage et méthodes d'accessibilité aux données*; Memory to obtain the PhD degree, École Polytechnique de Montréal: Canada, 2009.
17. Johnson, D. B.; Maltz, D. A. *Dynamic Source Routing in Ad Hoc Wireless Networks*; Computer Science Department Carnegie Mellon University: USA, 1996.
18. Lattab, Y.; Slaouti, O. *Optimisation par colonie de fourmis de la gestion de l'énergie appliquée au modèle MWAC*; Memory to obtain the engineering degree in informatics, École nationale Supérieure d'Informatique, 2010.
19. Li, Q.; Aslam, J.; Rus, D. In *Hierarchical Power-Aware Routing in Sensor Networks,* Proceedings of the DIMACS Workshop on Pervasive Networking, Rutgers University, Piscataway, NJ, May 2001.

20. Lindsey, S.; Raghavendra, C. S. In *PEGASIS: Power Efficient Gathering in Sensor Information Systems,* Proceedings of the IEEE Aerospace Conference, Big Sky, Montana, USA, March 2002.

21. Manjeshwar, A.; Agrawal, D. P. In *TEEN: A Protocol for Enhanced Efficiency in Wireless Sensor Networks,* Proceedings of the First International Workshop on Parallel and Distributed Computing Issues in Wireless Networks and Mobile Computing, San Francisco, CA, USA, April 2001.

22. Rodoplu, V.; Meng, T. H. Minimum Energy Mobile Wireless Networks. *IEEE J. Sel. Areas Commun.* **1999**, *17*(8), 1333–1344.

23. Sadagopan, N.; Krishnamachari, B.; Helmy, A. In *The ACQUIRE Mechanism for Efficient Querying in Sensor Networks,* Proceedings of the First International Workshop on Sensor Network Protocol and Applications, Alaska, USA, May, 2003.

24. Shah, R. C.; Rabaey, J. M. In *Energy Aware Routing for Low Energy Ad Hoc Sensor Networks,* Proceedings of the IEEE Wireless Communications and Networking Conference (WCNC), Orlando, FL, USA, March 2002.

25. Singh, S. K.; Singh, M. P.; Singh, D. K. Routing Protocols in Wireless Sensor Networks—A Survey. *Int. J. Comput. Sci. Eng. Surv. (IJCSES)* 2010,1(2), pp.63–83.

26. Sohrabi, K.; Pottie, J.; Gao, J.; Ailawadhi, V. Protocols for Self Organization of a Wireless Sensor Network. *IEEE Pers. Commun.* **2000**, *7*(5), 16–27.

27. Subramanian, L.; Katz, R. H. In *An Architecture for Building Self Configurable Systems,* Proceedings of IEEE/ACM Workshop on Mobile Ad Hoc Networking and Computing, Boston, MA, USA, August 2000.

28. Xu, Y.; Heidemann, J.; Estrin, D. In *Geography-Informed Energy Conservation for Ad Hoc Routing,* Proceedings of the 7th Annual ACM/IEEE International Conference on Mobile Computing and Networking, Rome, Italy, July 2001.

29. Ye, F.; Chen, A.; Liu, S.; Zhang, L. In *A Scalable Solution to Minimum Cost Forwarding in Large Sensor Networks,* Proceedings of the tenth International Conference on Computer Communications and Networks (ICCCN),USA, 2001.

30. Younis, O.; Fahmy, S. *Distributed Clustering in Ad-hoc Sensor Networks: A Hybrid, Energy-efficient Approach*; Research Report, Department of Computer Sciences, Purdue University: USA, September 2002.

31. Yu, Y. Estrin, D.; Govindan, R. *Geographical and Energy-Aware Routing: A Recursive Data Dissemination Protocol for Wireless Sensor Networks*; UCLA Computer Science Department Technical Report: LA, USA, May 2001.

32. Zhu, J.; Zhao, H.; Xu, J. *An Energy Balanced Reliable Routing Metric in WSNs*; The Northeastern University: Shen Tang, China, February 18, 2009.

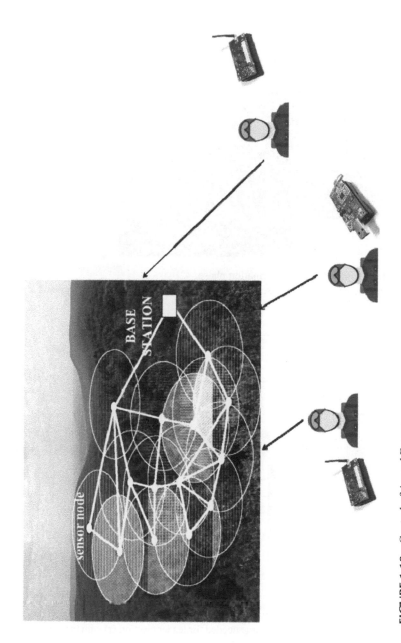

FIGURE 1.10   Control of Armed Forces.
Source: Adapted from ref 11.

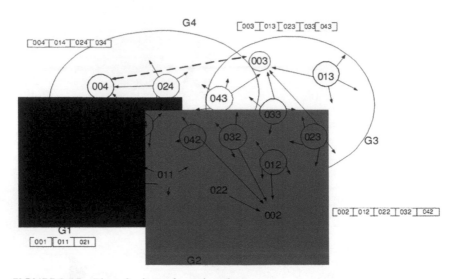

FIGURE 2.15 Phase 2, cluster formation phase.
Source: Reprinted from ref [3]. Used with permission of the Creative Commons License.
https://creativecommons.org/licenses/by/4.0/.

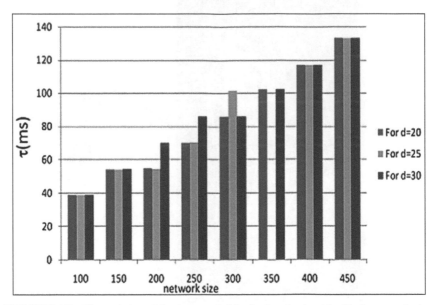

FIGURE 2.17 The variation of the time needed for leader election (τ) with the network size.
Source: Reprinted from ref [3]. Used with permission of the Creative Commons License.
https://creativecommons.org/licenses/by/4.0/.

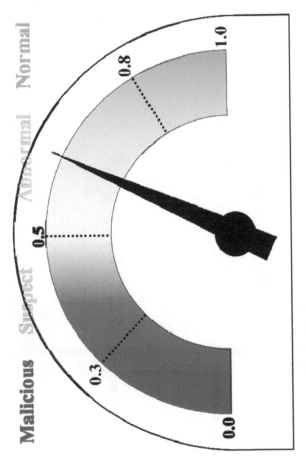

FIGURE 5.1 The behavior level (BL_i).

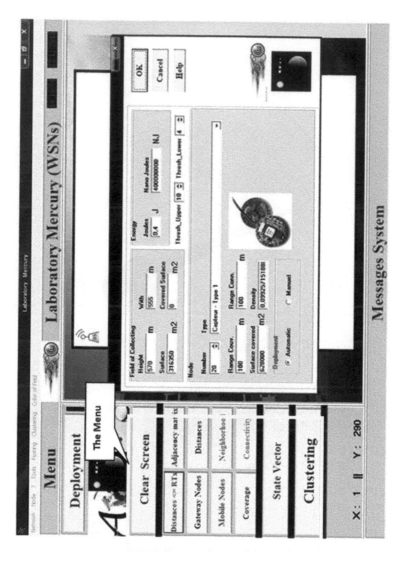

FIGURE 6.6 Help topic of our virtual laboratory mercury.

CHAPTER 5

A DISTRIBUTED AND SAFE WEIGHTED CLUSTERING ALGORITHM

5.1 INTRODUCTION

During the past few decades, clustering techniques have been extensively studied to improve the performance of WSNs[14]. In this section, we will review the existing clustering schemes which are most related to our work.

In the context of these surveyed research works about clustering in both ad-hoc networks and WSNs, The classification of our contribution among approaches was on the basis of computing the weight of each node in the network: the focus of our approach is put around strategy of distributed resolution and develops a new metric (the behavioral level metric) which promotes a safe choice of a cluster head in the sense where this last one will never be a malicious node. The monitor node's role is concerned by watching its neighbors to know what each one of them does with the messages it receives from another neighbor. The monitor fails if its neighbor do not retransmitted that message. Such importance was not given by those previous works to the election criteria of nodes that are responsible for monitoring the network.

5.2 A SURVEY OF CLUSTERING SCHEMES

In this section, we outline some approaches of clustering used in ad hoc networks and wireless sensor networks (WSNs). Research studies on clustering in ad hoc networks involve surveyed works on clustering algorithms[1–4] and cluster head (CH) election algorithms.[5,6] Abbasi and Younis[7] presented taxonomy and classification of typical clustering schemes, then summarized different clustering algorithms for WSNs based on classification of variable convergence time protocols and constant convergence time algorithms, and highlighted their objectives, features, complexity, and so forth.

A single metric based on clustering as in ref[8] shows that the node with the least stability value is elected as CH among its neighbors. However, the choice of CH which has a lower energy level could quickly become a bottleneck of its cluster. Er and Seah[9] designed and implemented a dynamic energy-efficient clustering algorithm for mobile ad hoc networks (MANETs) that increases the network lifetime. The proposed model elects first the nodes that have a higher energy and less mobility as CHs, then periodically monitors the CH's energy, and locally alters the clusters to reduce the energy consumption of the suffering CHs. The algorithm defines a yellow threshold to achieve some sort of local load balancing and a red threshold to trigger local re-clustering in the network. However, the cluster formation in this scheme is not based on connectivity, so the formed clusters are not well connected; consequently, this increases the re-affiliation rate and maximizes re-clustering situations. Jain and Reddy[10] have proposed a novel method of modeling wireless sensor network using fuzzy graph and energy-efficient fuzzy-based k-hop clustering algorithm which takes into account the dynamic nature of network, volatile aspects of radio links, and physical layer uncertainty. They have defined a new centrality metric, namely fuzzy k-hop centrality. The proposed centrality metric considers residual energy of individual nodes, link quality, hop distance between the prospective CH, and respective member nodes to ensure better CH selection and cluster quality, which results in better scalability, balancing of energy consumption of nodes, and longer network lifetime.

Other proposals use a strategy based on computed weight in order to elect CHs.[11,12,5,2] The main strategy of these algorithms is based mainly on adding more metrics such as the connectivity degree, mobility, residual energy, and the distance of a node from its neighbors, corresponding to some performance in the process of electing CHs. Although the algorithms which use this strategy allow us to ensure the election of better CHs based only on their high computed weight from the considered metrics, they unfortunately do not ensure that the elected CHs are legitimated nodes, that is, whether the election process of CHs is safe or not. Safa et al.[13] propose a novel cluster-based trust-aware routing protocol (CBTRP) for MANETs to protect forwarded packets from intermediary malicious nodes.

The proposed protocol ensures the passage of packets through trusted routes only by making nodes monitor the behavior of each other and update their trust tables accordingly. However, in CBTRP, all nodes monitor the network which lead to rapid drainage of node energy and therefore minimize the lifetime of the network. In the last chapter, we showed that WSNs are vulnerable to various types of attacks.[14,15] In the last decade, several studies proposed solutions to solve attacks in WSNs by using cryptography, such as SPINS[16]. However, cryptography alone is not enough to prevent node compromise attacks and novel misbehavior in WSNs[17]. Khalil et al.[18] propose a protocol called DICAS which uses local monitoring and mitigates the attacks against control traffic by detecting, diagnosing, and isolating the malicious nodes. Marti et al. [MGLB05] use a watchdog technique or local monitoring for ad hoc networks in order to improve the detection of mischievous nodes. They use a technique called path rater to help routing protocols to avoid them. A self-monitoring mechanism that pays more attention to the system-level fault diagnosis of the network was proposed by Hsin et al.[19], especially for detecting node failures. However, they did not deal with malicious behaviors. Little effort has been made to include the security aspect in the clustering mechanism. Yu et al.[20] try to secure the clustering mechanism against wormhole attack in ad hoc networks (communication between CHs). However, this is done after forming clusters, not during the election procedure of CHs. Liu[21] surveyed the clustering algorithms available for WSNs but that was done from the perspective of data routing. Hai et al.[22] propose a lightweight intrusion detection framework integrated for clustered sensor networks by using an overhearing mechanism to reduce the sending alert packets. Elhdhili et al.[23] propose a reputation-based clustering algorithm that aims to elect trustworthy, stable, and high-energy CHs but during the election procedure, not after forming clusters. Benahmed et al.[24] used clustering mechanism based on weighted computing as an efficient solution to detect misbehavior nodes during distributed monitoring process in WSNs. However, they focused only on the misbehavior of malicious nodes and not on the nature of attacks, the formed clusters are not homogeneous, the proposed algorithm SDCA is not coupled with a routing protocols, and it does not give much importance to energy consumption.

- Portions of this chapter are reprinted from (A. Dahane, A. Loukil, B. Kechar, N. Berrached, "Energy Efficient and Safe Weighted Clustering Algorithm for Mobile Wireless Sensor Networks," Mobile Information Systems, 2015) Adapted.
- Portions of this article are reprinted from (A. Dahane, N. Berrached, A. Loukil, "A Virtual Laboratory to Practice Mobile Wireless Sensor Networks: A Case Study on Energy Efficient and Safe Weighted Clustering Algorithm," Journal of Information Processing Systems (JIPS), Vol. 11, No. 2, pp. 205–228, June 2015) Modified.
- Portions of this article are reprinted from (A. Dahane, N. Berrached, A. Loukil, "Balanced and Safe Weighted Clustering Algorithm for Mobile Wireless Sensor Networks," The 5th International Conference on Computer Science and its Applications (CIIA'15), Chapter book, (Springer), Vol. 456, pp 429–441, May 20–21, SAIDA, Algeria, 2015) Modified.

5.3   CLUSTERING ALGORITHM FOR MOBILE WSNS

5.3.1   PRELIMINARIES

In this section, we first present some assumptions of the proposed algorithm "a distributed and safe weighted clustering algorithm (ES-WCA)." Then we present in detail an extended version of ES-WCA[25] followed by an illustrative example. Before discussing the technical details of the algorithm "ES-WCA," we present some definitions and notations that are necessary to understanding.

5.3.2   NETWORK MODELING

We consider a wireless network 1-hop where all nodes cooperate in order to ensure communications between them. Station base is responsible to recuperate the data and analyze it. Such a wireless network can be represented by a connected graph $G=(U, N)$ where U is the set of nodes (sensors), $U \subseteq N^2$. The set of arcs reflecting the possible direct communication between nodes:

"Oriented pair (i, j) belongs to N if and only if the node u can send a message to node v," and we say that: "v is adjacent to u" or "u also covers v."

Weight () is a function associating to each node $i \in N$ a weight $W(i)$ ∈ representing its capacity to be cluster head couples belonging to N and depending on the position of the nodes and their transmission range. We assume that all nodes have the same transmission range, and all links in the network are bidirectional, that is to say: if "i" is a neighbor of "j" then "j" is a neighbor of "j."

For each node u, we assign a single value which characterizes the called Identifier and recorded Id.

We used the model of the disk unit (unit disk graph).[26,27] This model is very responded to model communication between nodes in broadcast protocols designed for ad hoc and sensor networks. In this model, it is assumed that two nodes can communicate with each other if the Euclidean distance "dist (i, j)" between them is not greater than a given transmission range, and messages are always received without any errors.

Hence, the set N can be defined as follows (see Table 5.1):

$$N(i) = \left\{ n_i \ / \ dist(i, j) < tx_{range} \ with \ i \neq j \right\}$$

TABLE 5.1 Necessary Adjustments to Create a New Network.

Range of connectivity	It depends on the type of sensors
Cover range	It depends on the type of sensors
Surface to cover	$= \pi * (\text{Range of connectivity})^2 * \text{Number of nodes}$
Surface (field)	$= \text{Height} * \text{Width}$
Density	$= \text{Surface to cover}/(\text{Surface} * \text{Number of nodes})$

5.3.3 ASSUMPTIONS

This work is based on the following assumptions:

a) The network formed by the nodes and the links can be represented by an undirected graph $G = (U, E)$, where U represents the set of nodes n_i and E represents the set of links ei.[25, 28,,29]

b)   All sensor nodes are deployed randomly in a two-dimensional (2D) plane.

c)   A node interacts with its 1-hop neighbors directly and with other nodes through intermediate nodes using multi-hop packet forwarding based on a routing protocol such as ad hoc on demand distance vector[30,31,32,45] or destination-sequenced distance vector.[33,31,32,34]

d)   The radio coverage of sensor nodes is a circular region centered on this node with radius R.

e)   Two sensor nodes cannot be deployed in exactly the same position (x, y) in a 2D space.

f)   All sensor nodes are identical or homogeneous. For example, they have the same radio coverage radius R.

g)   Each node can determine its position at any moment in a 2D space.

h)   Each cluster is monitored by only one CH.

i)   Each cluster member (CM) communicates directly with its CH for the transmission of security metrics.

j)   A CH communicates directly with the base station for the transmission of security information and possible alerts.

5.4   METRICS FOR CHS ELECTION

This section introduces the different metrics used for CH election by focusing on behavior-level metric.

5.4.1   THE BEHAVIOR LEVEL OF NODE (BL$_i$)

The behavioral level of a node is a key metric in our contribution. Initially, each node is assigned an equal static behavior level "$BL_i = 1$." However, this level can be decreased by the anomaly detection algorithm if a node misbehaves.

For computing the behavior level of each node, nodes with a behavior level less than threshold behavior will not be accepted as CH candidates even if they have the other interesting characteristics such as high energy, high degree of connectivity, or low mobility. Nevertheless, abnormal nodes and suspect nodes may belong to a cluster as CM but never as CH. Therefore, we define the behavior level of each sensor node n_i, noted BL_i, in any neighborhood of the network with the mapping function (Eq. 5.1), as illustrated by Figure 5.1:

$$Mp(BL_i) = \begin{cases} \text{Normal node:} & 0.8 \le BL_i \le 1 \\ \text{Abnormal node:} & 0.5 \le BL_i < 0.8 \\ \text{Suspect node:} & 0.3 \le BL_i < 0.5 \\ \text{Malicious node:} & 0 \le BL_i < 0.3 \end{cases} \qquad (5.1)$$

The values in formula (Eq. 5.1) are chosen on the basis of several reputed models of WSNs adopted by numerous researchers like Shaikh et al.[35] and Lehsaini et al.[36]. The monitor node watches its neighbors to know what each one of them does with the messages it receives from another neighbor.

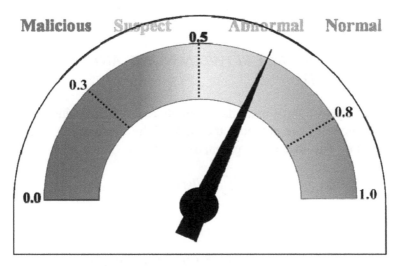

FIGURE 5.1 (See color insert.) The behavior level (BL_i).

If the neighbor of the monitor changes, delays, replicates, or simply keeps a message that should be retransmitted, the monitor counts a failure.

The number of failures has influence on the behavior of neighbors; for instance, if the monitor counts one failure from a neighbor, its behavior will decrease by 0.1 units. This allows the monitor (CH) to differentiate malicious nodes (that make much failure) of a legitimate node (that make fewer failure) in case there are collisions.

5.4.2 THE MOBILITY OF NODE (M_i)

Our objective is to have stable clusters. Therefore, we have to elect nodes with low relative mobility as CHs. To characterize the instantaneous nodal mobility, we use a simple heuristic mechanism as presented in the formula below (Eq. 5.2):[37,38,39,45]

$$M_i = \frac{1}{T} \sum_{t=1}^{T} \sqrt{(x_t - x_{t-1})^2 + (y_t - y_{t-1})^2} \tag{5.2}$$

where (x_t, y_t) and (x_{t-1}, y_{t-1}) are the coordinates of node at time t and $t-1$, respectively. T is the period for which this parameter is estimated.

5.4.3 THE DISTANCE BETWEEN NODE AND ITS NEIGHBORS (D_i)

This is likely to reduce node detachments and enhance cluster stability. For each node n_i, we compute the sum of the distances D_i with all its neighbors n_j. This distance is given, as in refs 28, 12, 42, 45 by (Eq. 5.3):

$$D_i = \sum_{j \in N(i)} \{dist(i,j)\} \tag{5.3}$$

5.4.4 THE RESIDUAL ENERGY OF NODE n_i (Er_i)

The residual energy of a node n_i, after transmitting a message of k bits at distance d from the receiver, is calculated according to refs 6,40,41 (Eq. 5.4):

$$Er_i = E - \left(E_{Tx}(k,d) + E_{Rxelec}(k) \right) \tag{5.4}$$

where
- E: the node's current energy.
- $(k, d) = k \cdot E_{elec} + k \cdot E_{amp} \cdot d^2$: it refers to the required energy to send a message, where E_{amp} is the required amplifier energy.
- $E_{Rxelec}(k) = k E_{elec}$: it refers to the energy consumed while receiving a message.

5.4.5 THE DEGREE OF CONNECTIVITY OF NODE n_i AT TIME t (C)

It represents the number of n_i's neighbors given by Equation 5.5, according to refs 23,25.

$$C_i = |N(i)| \qquad (5.5)$$

where
- dist (i, j): distance separating two nodes n_i and n_j
- tx_{range}: the transmission radius.

For each node, we must calculate its weight P_i, according to the equation (Eq. 5.6):

$$P_i = w_1 * BL_i + w_2 * Er_i + w_3 * M_i + w_4 * C_i + w_5 * D_i \qquad (5.6)$$

where w_1, w_2, w_3, w_4, and w_5 are the coefficients corresponding to the system criteria, so that (Eq. 5.7):

$$w_1 + w_2 + w_3 + w_4 + w_5 = 1 \qquad (5.7)$$

We propose to generate homogeneous clusters whose size lies between two thresholds: $Thresh_{Upper}$ and $Thresh_{Lower}$. These thresholds are arbitrarily selected or they depend on the topology of the network. Thus, if their values depend on the topology of the network, they are calculated according to refs 36,40,42,45 as follows:

i.: the node that has the maximum number of neighbors with one jump:

$$\delta_{12}(u) = \max\left(\delta_{12}(u_i): \quad u_i \in U\right) \qquad (5.8)$$

ii.: the node that has the minimal number of neighbors with one jump:

$$\delta_{12}(v) = \min\left(\delta_{12}(v_i): \quad v_i \in U\right) \tag{5.9}$$

We denote AVG by the average cardinal of the groups with one jump of all the nodes of the network, as below (Eq. 5.10):

$$AVG = i = 1n\delta_{12}u_iN \tag{5.10}$$

where N represents the number of nodes in the network. Thus, the two thresholds are calculated as follows (Eqs. 5.11 and 5.12):

$$Thresh_{Upper} = \frac{1}{2}\left(\delta_{12}(u) + AVG\right) \tag{5.11}$$

$$Thresh_{Lower} = \frac{1}{2}\left(\delta_{12}(v) + AVG\right) \tag{5.12}$$

The calculated weight for each sensor is based on the above parameters (BL_i, M_i, D_i, Er_i, and C_i). The values of coefficient should be chosen depending on the basis of the importance of each metric in considered WSNs applications. For instance, it is possible to assign a greater value to the metric BL_i compared to other metrics if we promote the safety aspect in the clustering mechanism. It is also possible to assign the same value for each coefficient in the case where all metrics are considered as having the same importance. An approach based on these weight types will enable us to build a self-organizing algorithm which forms a small number of homogenous clusters in size and radius by geographically grouping close nodes. The resulting weighted clustering algorithm reduces energy consumption and guaranties the choice of legitimate CHs.

In the proposed heuristic approaches for sensor networks based on clustering technology, the members of a cluster do not transmit their collected data directly to the base station but to their corresponding CH. As a result, cluster heads are responsible for coordinating the CMs, aggregating the captured data, and transmitting them to a remote base station, directly or through a mode of multi-hop transmission. Therefore, since the cluster heads receive more packets and consume more energy to transmit with a long reach, so they are those whose energy will be exhausted soon if elected clusters for a long period.

Therefore, a technique clustering should avoid fixed election of cluster heads, as they are constrained by energy and can quickly deplete their batteries because of their high use. Thus, it can cause bottlenecks in clusters and consequently trigger the process or re-election of cluster heads again. For this reason, another contribution we proposed for remedying this failure is to repeat the CH election process periodically at each lapse of time Δt to fairly distribute energy consumption among sensors during the life of the network. In this work, we assume that each sensor has an omnidirectional antenna by simply allowing it to cover all transmission sensors located in its neighboring, and that the sensors are deployed in 2D space.

We consider the sensors are stable for a reasonable period during execution of the clustering process, and each sensor has a generic weight and is able to evaluate it. The weight represents the ability of a sensor to be a CH: more weight means higher priority.

- Portions of this article are reprinted from (A. Dahane, A. Loukil, B. Kechar, N. Berrached, "Energy Efficient and Safe Weighted Clustering Algorithm for Mobile Wireless Sensor Networks," Mobile Information Systems, 2015) Adapted.
- Portions of this article are reprinted from (A. Dahane, N. Berrached, A. Loukil, "A Virtual Laboratory to Practice Mobile Wireless Sensor Networks: A Case Study on Energy Efficient and Safe Weighted Clustering Algorithm," Journal of Information Processing Systems (JIPS), Vol. 11, No. 2, pp. 205–228, June 2015) Modified.
- Portions of this article are reprinted from (A. Dahane, N. Berrached, A. Loukil, "Safety of Mobile Wireless Sensor Networks Based on Clustering Algorithm," International Journal of Wireless Networks and Broadband Technologies (IJWNBT), Vol. 5, No. 1, pp. 73–102, June 2016) Adapted.
- Portions of this article are reprinted from (A. Dahane, N. Berrached, B. Kechar, "Energy Efficient and Safe Weighted Clustering Algorithm for Mobile Wireless Sensor Networks," The 9th International Conference on Future Networks and Communications (FNC'14), Procedia Computer Science, (Elsevier), August 17–20, Niagara Falls, Ontario, Canada, Vol. 34, pp. 63–70, 2014) Modified.

5.5   WEIGHTED CLUSTERING ALGORITHM (ES-WCA)

In this section, we present in detail an extended version of ES-WCA[25,29] followed by an illustrative example.

5.5.1   PROPOSED ALGORITHM

The ES-WCA algorithm that we present below is based on the ideas proposed by Chatterjee et al.[28], Lehsaini et al.[36], and Zabian et al.[5], with modifications made for our application. This algorithm runs in three phases: the setup phase, the re-affiliation phase, and the monitoring phase. It is worth nothing that the data flow diagram of ES-WCA is given in Figure 5.2. ES-WCA combines each of the above system parameters with certain weighting factors chosen according to the system needs.

5.5.2   THE SETUP PHASE

ES-WCA uses three types of messages in the setup phase (see Algorithm 5.1).[25,29] The message "CHmsg" is sent in the network by the sensor node which has the greatest weigh. The second one is the "JOINmsg" message which is sent by the neighbor of CH if it wants to join this cluster. Finally, a CH must send a response "ACCEPTmsg" message as shown in Figure 5.3.

The node which has the greatest weight begins the procedure by broadcasting "CHmsg" message to their 1-hop neighbors to confirm its role as a leader of the cluster. The neighbors confirm their role as being member nodes by broadcasting a "JOINmsg" message. In the case when nodes have the same maximum weight, the CH is chosen by using the best parameters ordered by their importance. If all parameters of nodes are equal, the choice is random.

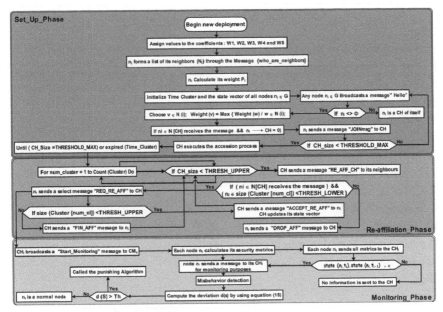

FIGURE 5.2   Flowchart of "ES-WCA."

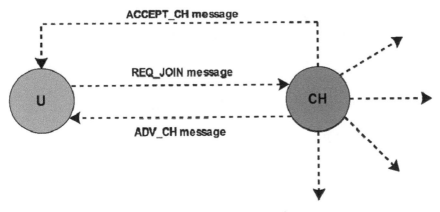

FIGURE 5.3   Procedure of affiliation of node "U" to a cluster.

Algorithm 5.1: Setup-phase algorithm

Begin
1: Assign values to the coefficients;
2: For any node **make**:
3: forms a list of its neighbours through the Message {who_are_neighbours};
4:;
5: Calculate its weight:
6:;
7: Initialize Time Cluster and the state vector of nodes Vector_State (Id, CH, Weight, List_Neighbors, Size, Nature)
8: CH = 0, Size = 0;
9: Nature =" None";
10: Repeat
11: Any node Broadcasts a message"Hello";
12: If Then
13: Choose;
14:;

15: The node that have the same maximum weight, the CH is the node that has the best criteria ordered by their importance (BL_i, Er_i, D_i and M_i) if all criteria of nodes are equal, the choice is random.

16: Else is a CH of itself.
EndIf
17: Update the state vector of the elected CH;
18: CH = ID;
19: Size = 1;
20: Nature = CH;
21: Send the message "CHmsg" by CH to its neighbours;
22: J = Count ();
23: For I = 1 to J **Do**
24: If (receives the message && $\rightarrow$ CH = 0)
25: Then sends a message "JOINmsg" to CH
26: If (CH $\rightarrow$ Size <)
27: Then CH sends a message "ACCEPTmsg" to Node;
28: CH executes the accession process;
29: CH $\rightarrow$ Size = CH $\rightarrow$ Size + 1;
30: executes the accession process;
31: $\rightarrow$ CH = CH $\rightarrow$ Id;
32: Else go to 10;

EndIf
EndIf
End For
33: Until expired (TimeCluster);
End.

- Portions of this chapter are reprinted from (A. Dahane, A. Loukil, B. Kechar, N. Berrached, "Energy Efficient and Safe Weighted Clustering Algorithm for Mobile Wireless Sensor Networks," Mobile Information Systems, 2015) Adapted.
- Portions of this article are reprinted from (A. Dahane, N. Berrached, A. Loukil, "A Virtual Laboratory to Practice Mobile Wireless Sensor Networks: A Case Study on Energy Efficient and Safe Weighted Clustering Algorithm," Journal of Information Processing Systems (JIPS), Vol. 11, No. 2, pp. 205–228, June 2015) Modified.
- Portions of this article are reprinted from (A. Dahane, N. Berrached, B. Kechar, "Energy Efficient and Safe Weighted Clustering Algorithm for Mobile Wireless Sensor Networks," The 9th International Conference on Future Networks and Communications (FNC'14), Procedia Computer Science, (Elsevier), August 17–20, Niagara Falls, Ontario, Canada, Vol. 34, pp. 63–70, 2014) Modified.

5.5.3 THE RE-AFFILIATION PHASE

ES-WCA uses four types of messages in the re-affiliation phase (see Algorithm 5.2).[40,42] The message "REAFFCH" is sent in the network by the CH whose cluster size is less than $Thresh_{Upper}$. These condone is the "REQREAFF" message which is sent by the neighbors of CH if it wants to join this cluster. Finally, a CH must send a response "ACCEPT_REAFF" message or "DROPAFF" message as illustrated by Figure 5.4.

Accordingly, in this phase, we propose to re-affiliate the sensor nodes belonging to clusters that have not attained the cluster size $Thresh_{Lower}$ to those that did not achieve $Thresh_{Upper}$ in order to reduce the number of clusters formed and organize them so as to obtain homogeneous and balanced clusters.

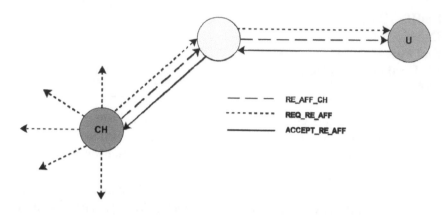

FIGURE 5.4   Procedure of re-affiliation of node "U" to a cluster.

Algorithm 5.2: Re-affiliation-phase algorithm

Inputs:,;
Outputs: set of clusters
Begin
1: Fornum_cl = 1 to Count (Cluster) **Do**
2: If (Size (Cluster [num_cl]) <)
Then
3: CH sends a message "RE_AFF_CH" to its neighbors
();
4: J = Count ();
EndIf
5: For I = 1 to J **Do**
6: If (receives the message)
&& ((Size (Cluster [num_cl]) <)
Then
7: sends a Select message "REQ_RE_AFF" to the CH;
8: If (Size (Cluster [num_cl]) <)
Then
9: CH sends a message "ACCEPT_RE_AFF" to;
10: CH updates its state vector;
11: CH → CH → Size = Size + 1;
12: updates its state vector;
13: → CH → ID = ID;
14: Else CH sends a "FIN_ AFF" message to;

15: Go to 2;
EndIF
16: Else sends a "DROP_AFF" message to CH;
EndIf
End For
End For
End.

- Portions of this chapter are reprinted from (A. Dahane, A. Loukil, B. Kechar, N. Berrached, "Energy Efficient and Safe Weighted Clustering Algorithm for Mobile Wireless Sensor Networks," Mobile Information Systems, 2015) Adapted.
- Portions of this article are reprinted from (A. Dahane, N. Berrached, A. Loukil, "A Virtual Laboratory to Practice Mobile Wireless Sensor Networks: A Case Study on Energy Efficient and Safe Weighted Clustering Algorithm," Journal of Information Processing Systems (JIPS), Vol. 11, No. 2, pp. 205–228, June 2015) Modified.
- Portions of this article are reprinted from (A. Dahane, N. Berrached, A. Loukil, "Balanced and Safe Weighted Clustering Algorithm for Mobile Wireless Sensor Networks," The 5th International Conference on Computer Science and its Applications (CIIA'15), Chapter book, (Springer), Vol. 456, pp 429–441, May 20–21, SAIDA, Algeria, 2015) Modified.

With the help of four figures (Figs. 5.5–5.8), our algorithm setup phase is demonstrated. Table 5.2 shows the quantitative results of the different criteria applied on the normal nodes ($BL_i \geq 0.8$).

Table 5.3 shows the weights P_i of neighbors for each node which has behavior BL_i higher than 0.8. The circles in Figure 5.5 represent the nodes, their identity Ids at the top, and their levels of behavior at the bottom. According to Table 5.3, node 1 could be attached to either CH11 or CH8 (since they have the same weight). However, the behavior level of node 11 is greater than that of node 8 (BL11 > BL8). So, node 1 will be attached to CH11.

TABLE 5.2 Weights of Neighbors.

Ids	BL_i	Er_i	C_i	D_i	M_i	P_i
1	0.86	3842.12	3	1.15	1.20	769.632
4	0.81	4832.54	5	2.30	0.30	968.133
5	0.88	4053.25	3	1.30	0.55	811.829
6	0.85	4620.43	0	0.00	0.20	924.361
8	0.81	4816.80	3	1.05	1.40	964.753
10	0.95	3650.25	2	0.55	0.10	730.805
11	0.91	4819.60	1	0.70	2.20	964.753

TABLE 5.3 Values of the Various Criteria of Normal Node.

Ids	1	4	5	6	8	10	11
1	769.632				964.753		964.753
4		968.133	811.829		964.753		
5		968.133	811.829			730.805	
6				924.361			
8	769.632				964.753		
10		968.133	811.829			730.805	
11	769.632						964.753

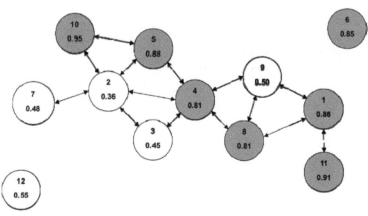

FIGURE 5.5 Topology of the network.

For the other nodes, we have various conditions. Node 4 declares itself as a CH. Node 5 will be attached to CH4. Node 6 declares itself as a CH, because it is an isolated node. Node 8 will be attached to CH4. Node 10 is connected to CH5, but Node 5 is attached to CH4. Thus, node 10 declares itself as a CH. Node 11 declares itself as a CH. These results give us the representation shown in Figure 5.6. Node 2 is connected to CH4 and CH10. Node 2 will be attached to CH4, because CH4 has the maximum weight (968.133). Node 3 is connected to CH4, which implies that node 3 will be attached to CH4. Node 7 is not connected to any CH, so node 7 declares itself as CH.

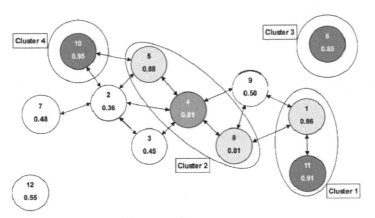

FIGURE 5.6   Identification of clusters node.

Node 9 is connected to CH4, and the node 9 will be attached to CH4. Node 12 is not connected to any CH, which implies that node 12 declares itself as a CH. These results give us the representation shown in Figure 5.7. We propose to generate homogeneous clusters whose size lies between two thresholds: $Thresh_{Upper} = 9$ and $Thresh_{Lower} = 6$. For that, we suggest to re-affiliate the sensor nodes belonging to the clusters that have not attained the cluster size $Thresh_{Lower}$ to those that did not reach $Thresh_{Upper}$. Node 4 has the highest weight and his size is less than $Thresh_{Upper}$. Nodes 1, 7, and 10 are neighbors of node 4 with 2 hops and belong to the clusters that have not attained the cluster size $Thresh_{Lower}$, so these nodes get merged to cluster 2.

Clusters 1, 3, and 4 will be homogeneous with cluster 1 when the network becomes densely. At the end of this example, we obtain a network of four clusters (as shown in Figure 5.8).

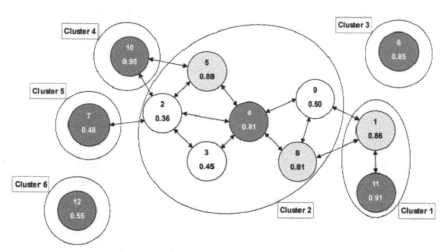

FIGURE 5.7   The final identification of clusters.

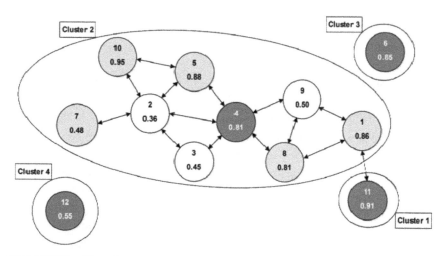

FIGURE 5.8   Final cluster structure (re-affiliation phase).

There are five situations that require the maintenance of clusters:

    i.   Battery depletion of a node,
    i.   Behavior level of a node less than or equal 0.3,
    ii.   Adding, moving, or deleting a node.

In all of these cases, if a node n_i is CH, then the setup phase will be repeated.

5.5.4 THE MONITORING PHASE

Monitoring in WSNs can be local with respect to a node or global with respect to the network, but in sensor networks, local monitoring is insufficient for detecting some types of errors and security anomalies. For this reason, in this work, we adopt a hybrid approach that is global monitoring based on distributed local monitoring (see Algorithm. 5.3).

The general architecture of our approach is illustrated in Figure 5.9. Our system "Mercury" detects the internal misbehavior nodes during distributed monitoring process in WSNs by the follow-up of the messages exchanged between the nodes. It is supposed that the network has already a mechanism of prevention to avoid the external attacks. All the received messages are analyzed by using a set of rules. A similar approach is followed by Da Silva et al.[43] and Benahmed et al.[44]. The monitoring process involves a series of steps,[29,39,45] as illustrated by the flowchart in Figure 5.10.

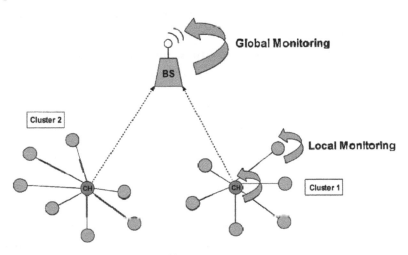

FIGURE 5.9   Monitoring-phase architecture.

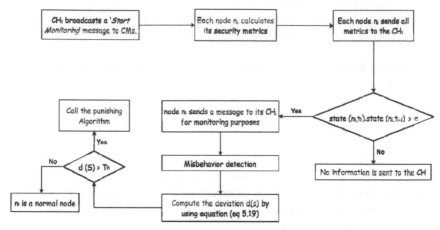

FIGURE 5.10   Monitoring phase.

Algorithm 5.3: Monitoring-phase algorithm

Step1: This step runs in each CH_i:

Each CH_i becomes the monitor node of its cluster members and broadcasts a "Start Monitoring" message with its Id_i to its entire cluster members (CMs).

Step2: Calculation of security metrics performed by each member of the cluster i.

Each node (i<>j) receives the message 'start monitoring' and calculates its security metrics as follows:

– Number of packets sent byat time interval $\Delta t = [t_0, t]$:

$$Nbp\_Send\left(ni, \Delta t\right)$$

– Number of packets received by node at time interval $\Delta t = [t_0, t_0]$:

$$Nbp\_Received\left(n_i, \Delta t\right)$$

– Delay between the arrivals of two consecutive packets:

$$Delay\_BP\left(n_i, t\right) = Arrival\_PT_i - Arrival\_PT_{i-} \quad 5.13 \quad (5.13)$$

– Energy consumption: The energy consumed by the node j in receiving and sending packets is measured using the following equation:

$$Ec(n_i, \Delta t) = Er(n_i, t_0) - Er(n_i, t_1) \qquad (5.14)$$

Where Δt is the time interval $[t_0, t_1]$; $Er(n_i, t_0)$ is the residual energy of node n_i as time t_0; $Er(n_i, t_1)$ is the residual energy of node n_i at time t_1 and $EC(n_i, \Delta t)$ is the energy consumption of node n_i at time interval.Δ

Step3: Sending all metrics to the CH.

After each consumption of the security metrics, the state of a node n_i at time t is denoted as state (n_i, t_i). For economy of storage volume, each node keeps only the latest state calculation.

In the initial deployment, each CM in cluster I sends some states (state n_i, t_i)) to the CHi for making a normal behavior model of node n_i by using a learning mechanism.

Each state contains the following information:

$$(Id, Nbp\_Send(ni, \Delta t), Nbp\_Received(n_i, \Delta t), Delay\_BP(n_i, t), Ec(n_i, \Delta t))$$

If (state n_i, t_i) – state $(n_i, t_{i-1}) \neq \in$ as a given threshold)

then node n_i sends a message $Msg = (Id, Nbp\_Send(ni, \Delta t), Nbp\_Received(n_i, \Delta t),$ $Delay\_BP(n_i, t), Ec(n_i, \Delta t))$ to its CH_i for monitoring purposes.

Otherwise no information is sent to the CH.

- The message received by CHiwill be stored in a table Tmet for future analysis.
- If a sensor node n_i does not respond during this monitoring period, it will be considered as misbehaving.
- The behavior level of sensor node n_i is computed using the following equation:

$$BL_i = BL_i - \text{rate} \qquad (5.15)$$

The rate is fixed on the basis of the nature of the application, for example, whether it is fault tolerant or not.

In our case, we took rate=0.1.

Step4: Misbehavior detection, which is performed by CH_i.

For each node n_i in the cluster i. the state in time slot t is expressed by the three-dimensional vector $S = (S_{t1}, S_{t2}, S_{t3})$.

Where

S_{t1} is the number of packets dropped by n_i as:

$$S_{t1} = \sum Ps_{Received\,by\,n_i} - \sum Ps_{Sent\,by\,n_i} - \sum Ps_{destined\,by\,n_i} \qquad (5.16)$$

With:

$$\sum Ps_{Received\,by\,n_i} = \sum Ps_{Sent\,by\,n_i} + \sum Ps_{destined\,by\,n_i} + \sum Ps_{lost\,by\,n_i} \qquad (5.17)$$

For a normal node: $S_{t1} \approx 0$;

- S_{t2} is the delay between the arrival of two consecutive packets, so that $S_{t2} = Delay\_BP(n_i, t)$.
- S_{t3} Is the energy consumption, so that. $S_{t3} = Ec(n_i, \Delta t)$

Here, $t \in [t_0, t_1] = \Delta t$;

In our case, the first interval is used for the training data set of n time slots. We calculate the mean vector $\bar{s}$ from S by using equation (Eq. 5.18).

$$\bar{S} = \frac{\sum_{t=t_0}^{t_n} S_t}{2} \qquad (5.18)$$

After modeling a normal behavior model for each sensor node, the behaviors of all nodes are sent to the base station for further analysis.

We compute the deviation d by using Equation (Eq. 5.19).

$$d(S) = \left| s - \bar{s} \right| \qquad (5.19)$$

When the distance is larger than threshold T_h (which means that it is out of the range of normal behavior), it will be judged as a misbehaving node (see Eq. 5.20).

In this case, the level of behavior is. $BL_i \approx 0$

$$\left\{ \begin{array}{l} d(S) > T_h : n_i \text{is an abnormal node, we will call the punishing algorithm} \\ \\ d(S) \leq T_h : n_i \text{is a normal node.} \end{array} \right\} \qquad (5.20)$$

The monitoring process involves a series of steps as illustrated by the flowchart Figure 5.10. In what follows, we present the punishing algorithm (see Algorithm. 5.4):

Algorithm 5.4: Punishing algorithm
Begin
1: I:=0;

2: I: = I+1;
3: If   ((I = Rate) && (<=0.1))
//Rate: parameter of maximum number of faults
defined by the administrator
4: then = - Rate;
5://Classification of the node according to its
6: If (BL_i =0.3) **then**
7: If (n_1 is CM) **then**
8: Suppression of the node of the list of the members;
9: Addition of the node to the blacklist;
EndIf
10: If (n_1 is CH) **then**//CH: Cluster Head
11: Addition of the node to the blacklist;
12: Set up Phase;
EndIf
EndIf
EndIf
End.

5.6   MODELING WITH UML

Modeling is a central part of all the activities that lead up to the deployment of good software. It helps to build models to communicate the desired structure and behavior of our system. For building models to better understand the system, we are designing and exposing opportunities for simplification, and reusing and managing risks.[46]

The Unified Modeling Language (UML) is a standard language for writing software blue prints. The UML may be used to visualize, specify, construct, and document the artifacts of a software-intensive system.

UML is defined as a graphical and textual modeling language intended to describe and understand needs, specify, develop solutions, and communicate viewpoints. It also unifies the notations required for various activities of a development process and provides, thereby, the means of establishing the monitoring of taken decisions, from requirements definition to coding. Here is a quick overview of the important UML diagrams that will be used throughout the project.[45] The UML was born from the merger of three methods that have most influenced object modeling in the middle of the 1990s.

1. Method OMT of Rumbaugh,
2. Method BOOCH '93 of Booch,
3. Method OOSE of Jacobson (Objet-Oriented Software Engineering) [GRA 09].

5.6.1   A MODEL

The modeling is to create a simplified representation of a problem: the model.

With the model, it is possible to simply be a concept and simulate it. The model has two components:

- The analysis, that is to say, the study of the problem.
- The design or the development of a solution of the problem.

The model is one possible representation of the system for a given viewpoint.

5.6.2 UML MODELING

UML provides a range of tools for representing all the object elements of the world (classes, objects, etc.) and the links between them. However, since only one representation is too subjective, UML provides a clever way to represent different projections of the same representation because of the views.[47]

A view is composed of one or several diagrams. There are two types of views:

- Static views
- Dynamic views
    - The use case diagram: It shows the structure of the necessary features for system users. It is normally used during the steps of capturing functional and technical requirements (see Fig. 5.11).
    - The class diagram: It is surely one of the most important diagrams in object-oriented development. On the functional side, this graph is intended to develop the structure of the entities

manipulated by users. At designing side, the class diagram shows the structure of object-oriented code (see Fig. 5.12).

– The sequence diagram: It represents the exchange of messages between objects, in the context of a particular operation of the system, (see Figs. 5.13, 5.14, and 5.15).

– The activity diagram: It represents the rules of sequence of activities and actions in the system. It can be treated as an algorithm but schematically, it addresses the dynamic view of a system.

– The object diagram: An object diagram shows a set of objects and their relationships. It represents static snapshots of instances of the things found in class diagrams. This diagram addresses the static design view of a system as do class diagrams [47] (see Fig. 5.16).

5.6.3   PROCESS OF MODELING WITH UML

5.6.3.1   PRELIMINARY STUDY

The preliminary study is the first step of the process. It consists to perform an initial identification of functional and operational requirements, mainly using text or simple diagrams. It prepares more formal activities of functional requirements capture and techniques capture.

5.6.3.2   PRESENTATION OF THE PROJECT TO ACHIEVE

It is an application that manages the wireless sensor network. It should help monitor the sink and the sensors constituting the network.

5.6.3.3   COLLECTION OF FUNCTIONAL REQUIREMENTS

This phase corresponds to a search which has enabled us to establish the following preliminary specification:

• A collection field consists of a set of sensors and sinks.

- The deployment is a random scatter of a set of nodes where the coverage of the collection field should be total or almost.
- The lifetime of network is submitted in the life and activity of deployed sensor at the collection field.

5.6.3.4   TECHNICAL OPTIONS

Here are the technical choices that have been adopted for the project:

- Modeling with UML
- Use of C Borland Language
- Using Cadifra release 1.3.1
- Identification of actors

We will list the actors likely to interact with the system, but first we give a definition of what an actor is.

Definition: an actor represents the abstraction of a role played by external entities (user, hardware device, or other system) that interact directly with the studied system. Actors in the system, initially identified, are:

- Network administrator (learner): An administrator can deploy the nodes, add and delete nodes, and give the order to the sink to activate the nodes.
- Sink: The sink can activate the nodes, communicates with administrator, and forms the routing table.

Sensor: The sensor may detect, send, and receive data.

Use case scenarios:

Title	Create a new network
Summary	This use case allows the learner to create a new network such as: it must specify the height and width of the surface field. The virtual laboratory "Mercury" offers the possibility to the learners to select a type of sensor from five predefined types (see Section 5.5). Each type has its characteristics (radius, energy, etc.). The student can also introduce his/her own characteristics.
Actors	Administrator (principal actor), the well field and the sensor (secondary actors).

Preconditions:

- Launch application "Clustering.exe."

Title	Adding a sensor
Summary	This use case allows the learner to add a sensor in the well field.
Actors	Administrator (principal actor), the well field and the sensor (secondary actors).

Preconditions:

- The surface field is empty or already contains sensors.

Title	Adjacency matrix
Summary	The administrator (learner) of the system seeking the display from the adjacency matrix of the graph $G = (U, N)$. Its elements are binary 1 or 0. 1: indicates that the two nodes are connected by a communication link before the clustering step (blue color). 0: means that the two nodes are not connected by a communication link before the clustering phase (blue color).
Actors	Administrator (principal actor), the well field and the sensor (secondary actors).

Preconditions:

Title	Gateway nodes
Summary	The administrator (learner) of the system seeks to detect gateway nodes of the graph $G = (U, N)$.
Actors	Administrator (principal actor), the well field and the sensor (secondary actors).

The well field contains sensors.

Preconditions:

- The well field contains sensors.
- The administrator partitioned the network into a number of clusters.
- There is a relationship between two clusters such as:
  - There is a neighboring node n_i of two CH or more CHs.
  - There is a node n_i neighbor of another node n_j or n_i and n_j are members of two different clusters.

Title	Clustering
Summary	This use case allows the administrator (learner) partitioning the network in to a number of clusters. The message "CHmsg" is sent in the network by the sensor node which has the greatest weigh. The second one is the "JOINmsg" message which is sent by the neighbor of CH, if it wants to join this cluster. Finally, a CH must send a response "ACCEPTmsg" message.
Actors	Administrator (principal actor), the well field and the sensor (secondary actors).

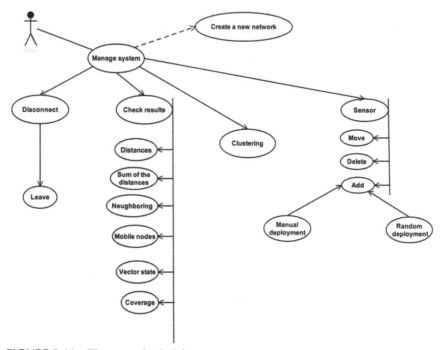

FIGURE 5.11   The network administrator use cases.

We are interested in modeling the functional aspect of a system, that is to say, we model what the system should do without saying how he will do it. A use case is a specific way of using the system. He used to describe what the system should do, without specifying how it will. All use cases should describe exhaustively the system requirements.

The class diagram:

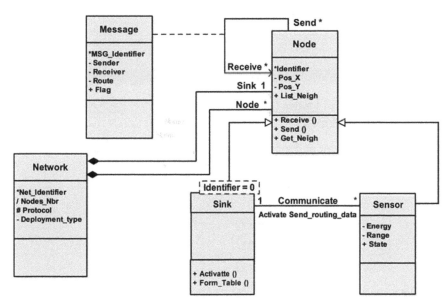

FIGURE 5.12 The class diagram.

The use case diagram:

Formalism:

Actors:

An actor is a user, human or not, the system in our case (Learner). In UML, an actor is represented graphically in three different ways (a, b, c):

a) In the form of an icon (stick man);
a) In a rectangular shape with the keyword "actor";
b) In the form intermediate between the two previous.

The sequence diagram:

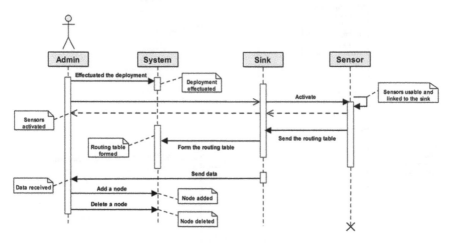

FIGURE 5.13   The network administrator sequence diagram.

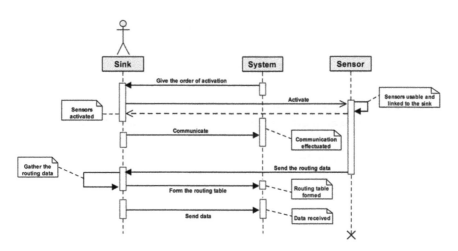

FIGURE 5.14   The sink sequence diagram.

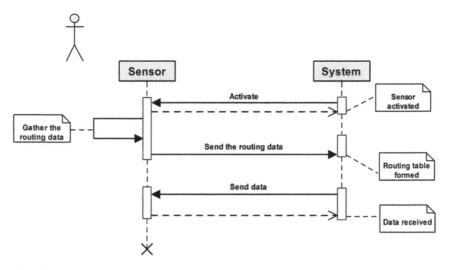

FIGURE 5.15 The sensor sequence diagram.

State transitions diagram:

UML state transition diagram uses the concept of finite state machines to be interested in the life cycle of an instance of a given class and its interactions with the rest of the system and in all possible cases (see Figure 5.17).

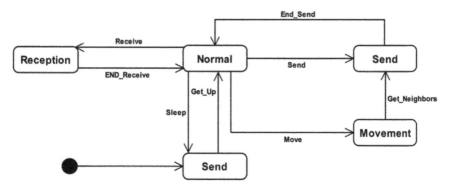

FIGURE 5.16 The objects diagram.

The objects diagram:

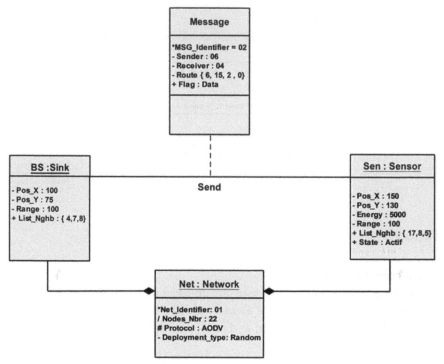

FIGURE 5.17   The sensor state diagram.

5.7 CONCLUSION

In this chapter, we proposed a new algorithm called ES-WCA for self-organization of mobile wireless sensor networks. "ES-WCA" takes into account the specifications and constraints of sensor networks and is suited to a particular type of sensor networks: mobile sensor networks. We set out with ES-WCA creating a stable virtual topology that minimize the frequent re-election and avoid the overall restructuring of the entire network. For this, we involved stability metric to elect the cluster head: the behavior level of node n_i (BL$_i$), the mobility of node ni (M_i), the distance between node n_i and its neighbors (D_i), the residual energy of node n_i (Er_i), and the degree of connectivity of node n_i at time t (C_i).

For each node, we must calculate its weight P_i, according to the Equation 5.6, where w_1, w_2, w_3, w_4, and w_5 are the coefficients corresponding to the system criteria, so that: $w_1+w_2+w_3+w_4+w_5=1$.

In the next chapter, we present the implementation of our system, results, and discussion.

Therefore, we propose to design our own simulator "Mercury." It is based on an object-oriented design and a distributed approach, such as a self-organization mechanism, which is distributed at the level of each sensor. The platform presented allowing students to make practical work and aims to show the feasibility, the flexibility, and the reduced cost of such a realization.

KEYWORDS

- **cluster head**
- **wireless sensor networks**
- **cluster-based trust-aware routing protocol**
- **weighted clustering algorithm**
- **cluster member**

REFERENCES

1. Chawla, M.; Singhai, J.; Rana, J. L. Clustering in Mobile Ad Hoc Networks: a Review. *Int. J. Comput. Sci. Inf. Secur.* **2010**, *8*(2), 293–301.
2. Agarwal, R.; Gupta, R.; Motwani, M. Review of Weighted Clustering Algorithms for Mobile Ad Hoc Networks. *Comput. Sci. Telecommun.* **2012**, *33*(1), 71–78.
3. Yu, J. Y.; Chong, P. H. J. A Survey of Clustering Schemes for Mobile Ad Hoc Networks. *IEEE Commun. Surv. Tutor.* **2005**, *7*(1), 32–48.
4. Sun, B.; Gui, C.; Song, Y.; Hu, C. Stable Cluster Head Selection Algorithm for Ad Hoc Networks. *Int. J. Future Gener. Commun. Networking* **2013**, *6*(3), 95–105.
5. Zabian, A.; Ibrahim, A.; Al-Kalani, F. Dynamic Head Cluster Election Algorithm for Clustered Ad-Hoc Networks. *J. Comput. Sci.* **2008**, *4*(1), 42–50.
6. Soro, S.; Heinzelman, W. B. Cluster Head Election Techniques for Coverage Preservation in Wireless Sensor Networks. *Ad Hoc Netw.* **2009**, *7*(5), 955–972.
7. Abbasi, A. A.; Younis, M. A Survey on Clustering Algorithms for Wireless Sensor Networks. *Comput. Commun.* **2007**, *30*(14), 2826–2841.
8. Ekici, E.; Gu, Y.; Bozdag, D. Mobility-Based Communication in Wireless Sensor Networks. *IEEE Commun. Mag.* **2006**, *44*(7), 56–62.

9. Er II.; Seah, W. K. G. In *Mobility-Based d-Hop Clustering Algorithm for Mobile Ad Hoc Networks*. Proceedings of IEEE Wireless Communications and Networking Conference (WCNC), Atlanta, GA, 2004, 2359–2364.

10. Jain, A.; Reddy, B. V. R. A Novel Method of Modeling Wireless Sensor Network Using Fuzzy Graph and Energy Efficient Fuzzy Based K-Hop Clustering Algorithm. *Wirel. Pers. Commun.* **2015**, *82*(1), 157–181.

11. Chatterjee, M.; Das, S. K.; Turgut, D. Wca: a Weighted Clustering Algorithm for Mobile Ad Hoc Networks. *Clust. Comput.* **2002**, *5*(2), 193–204.

12. Choi, W.; Woo, M. In *A Distributed Weighted Clustering Algorithm for Mobile Ad Hoc Networks*. Proceedings of Advanced International Conference on Telecommunications and International Conference on Internet and Web Applications and Services (AICT-ICIW '06), Guadeloupe, French Caribbean, 2006, 73–73.

13. Safa, H.; Artail, H.; Tabet, D. A Cluster-Based Trust-Aware Routing Protocol for Mobile Ad Hoc Networks. *Wirel. Netw.* **2010**, *16*(4), 969–984.

14. Dargie, W.; Poellabauer, C. *Fundamentals of Wireless Sensor Networks: Theory and Practices*. Wiley Series on Wireless Communications and Mobile Computing, **2010**.

15. Kavita, T.; Sridharan, D. Security Vulnerabilities in Wireless Sensor Networks: a Survey. *J. Inf. Assur. Secur.* **2010**, *5*, 31–44.

16. Perrig, A.; Szewczyk, R.; Tygar, J. D.; Wen, V.; Culler, D. E. Spins: Security Protocols for Sensor Networks. *Wirel. Netw.* **2002**, *8*(5), 521–534.

17. Ganeriwal, S.; Srivastava, M. B. In *Reputation-Based Framework for High Integrity Sensor Networks*. Proceedings of the 2nd ACM Workshop on Security of Ad Hoc and Sensor Network (SASN 2004), Washington, DC, 2004; pp 66–77.

18. Khalil, I.; Bagchi, S.; Shroff, N. B. In *LITEWORP: A Lightweight Countermeasure for the Wormhole Attack in Multihop Wireless Networks*. Proceedings of International Conference on Dependable Systems and Networks (DSN 2005), Yokohama, Japan, 2005, 612–621.

19. Hsin, C.; Liu, M. Self-Monitoring of Wireless Sensor Networks. *Comput. Commun.* 2006, 29(4), 462–476.

20. Yu, Y.; Zhang, L. In *A Secure Clustering Algorithm in Mobile Ad Hoc Networks*. Proceedings of IACSIT Hong Kong Conferences (IPCSIT Vol. 29), 2012; pp 73–77.

21. Liu, X. A Survey on Clustering Routing Protocols in Wireless Sensor Networks. *Sensors* **2012**, *12*(8), 11113–11153.

22. Hai, T. H.; Huh, E. N.; Jo, M. A Lightweight Intrusion Detection Framework for Wireless Sensor Networks. *Wirel. Commun. Mob. Comput.* **2010**, *10*(4), 559–572.

23. Elhdhili, M. E.; Azzouz, L. B.; Kamoun, F. Reputation Based Clustering Algorithm for Security Management in Ad Hoc Networks with Liars. *Int. J. Inf. Comput. Secur.* **2009**, *3*(3–4), 228–244.

24. Benahmed, K.; Haffaf, H.; Merabti, M. Monitoring of Wireless Sensor Networks. In *Sustainable Wireless Sensor Networks*; Seah, W., Tan, Y. K., Eds.; InTech: Rijeka, Croatia, 2010.

25. Dahane, A.; Berrached, N.; Kechar, B. In *Energy Efficient and Safe Weighted Clustering Algorithm for Mobile Wireless Sensor Networks*. The 9th International Conference on Future Networks and Communications (FNC'14), Procedia Computer Science, (Elsevier), August 17–20, Niagara Falls, Ontario, Canada 2014; Vol. 34, pp 63–70.

26. Lehsaini M, Diffusion et couverture basées sur le clustering dans les réseaux de capteurs: application à la domotique (Thèse de doctorat), 2009.

27. Clark, B. N.; Colbourn, C. J.; Johnson, D. S. Unit Disk Graphs. Discrete Math. 1990, 86, 165–177.

28. Chatterjee, M.; Das, S. K.; Turgut, D. Wca: A Weighted Clustering Algorithm for Mobile Ad Hoc Networks. *Clust. Comput.* 2002, 5(2), 193–204.

29. Dahane, A.; Loukil, A.; Kechar, B.; Berrached, N. Energy Efficient and Safe Weighted Clustering Algorithm for Mobile Wireless Sensor Networks. Mobile Information Systems, 2015.

30. Taneja, S.; Kush, A. A Survey of Routing Protocols in Mobile Ad Hoc Networks. *Int. J. Innov. Manage. Technol.* 2010, 1(3), 279–285.

31. Dahane, A.; Berrached, N.; Hibi, A.; Loukil, A. *Routing in Wireless Sensor Networks a Comparative Study: Between AODV and DSDV.* International Conference on Embedded Systems in Telecommunications and Instrumentation (ICESTI '14), 2014.

32. Dahane, A.; Berrached, N.; Loukil, A. A Virtual Laboratory to Practice Mobile Wireless Sensor Networks: a Case Study on Energy Efficient and Safe Weighted Clustering Algorithm. *J. Inf. Process Syst.* 2015, 11(2), 205–228.

33. Perkins, C. E.; Bhagwat, P. Highly Dynamic Destination-Sequenced Distance-Vector Routing (DSDV) for Mobile Computers. *ACM SIGCOMM Comput. Commun. Rev.* 1994, 24(4), 234–244.

34. Dong, Q.; Dargie, W. A Survey on Mobility and Mobility Aware Mac Protocols in Wireless Sensor Networks. *IEEE Commun. Surv. Tutor.* 2011, 15(1), 88–100.

35. Shaikh, R. A.; Jameel, H.; Lee, S.; Rajput, S.; Song, Y. J. In *Trust Management Problem in Distributed Wireless Sensor Networks.* Proceedings of 12th IEEE International Conference on Embedded and Real-Time Computing Systems and Applications, Sydney, 2006; pp 411–414.

36. Lehsaini, M.; Guyennet, H.; Feham, M. An Efficient Cluster-Based Self-Organisation Algorithm for Wireless Sensor Networks. *Int. J. Sens. Netw.* 2010, 7(1), 85–94.

37. Hussein, A. H.; Abu Salem, A. O.; Yousef, S. In *A Flexible Weighted Clustering Algorithm Based on Battery Power for Mobile Ad Hoc Networks.* Proceedings of IEEE International Symposium on Industrial Electronics (ISIE 2008), Cambridge, UK, 2008, 2102–2107.

38. Li, C.; Wang, Y.; Huang, F.; Yang, D. In *A Novel Enhanced Weighted Clustering Algorithm for Mobile Networks.* Proceedings of the 5th International Conference on Wireless Communications, Networking and Mobile Computing (WiCom '09), Beijing, China, 2009; pp 1–4.

39. Dahane, A.; Berrached, N.; Loukil, A. In *Energy Efficient and Safe Weighted Clustering Algorithm for Mobile Wireless Sensor Networks.* The 6th International Conference on Ambient Systems, Networks and Technologies (ANT'15), Procedia Computer Science, (Elsevier), June 02–05, London, UK, 2015; Vol. 52, pp 641–646.

40. Graine, S.; 2 pour une modélisation orientée objet UML, Exercices Corrigés, Les éditions L'abeille, 2009.

41. Dahane, A.; Berrached, N.; Loukil, A. Balanced and Safe Weighted Clustering Algorithm for Mobile Wireless Sensor Networks. The 5th International Conference on Computer Science and its Applications (CIIA'15), Chapter book, (Springer), Vol. 456, pp 429–441, May 20–21, 2015, SAIDA, Algeria.

42. Dahane, A.; Berrached, N.; Loukil, A. In *Homogenous and Secure Weighted Clustering Algorithm for Mobile Wireless Sensor Networks*. The 3rd International Conference on Control Engineering and Information Technology (CEIT '15), IEEE, May 25–27, 2015, Tlemcen, Algeria.

43. Dahane, A.; Berrached, N.; Loukil, A. Safety of Mobile Wireless Sensor Networks Based on Clustering Algorithm. *Int. J. Wirel. Netw. Broadband Technol. (IJWNBT)*, **2016**, *5*(1), 73–102, (June).

44. da Silva, A. P. R.; Martins, M. H.; Rocha, B. P.; Loureiro, A. A.; Ruiz, L. B.; Wong, H. C. In *Decentralized Intrusion Detection in Wireless Sensor Networks*. Proceedings of the 1st ACM International Workshop on Quality of Service and Security in Wireless and Mobile Networks, Montreal, Canada, 2005; pp 16–23.

45. Benahmed, K.; Merabti, M.; Haffaf, H. Distributed Monitoring for Misbehaviour Detection in Wireless Sensor Networks. *Secur. Commun. Netw.* **2013**, *6*(4), 388–400.

46. Dino, G. *The Unified Modeling Language User Guide*, 1st ed. Networking Laboratory Helsinki, Addison Wesley: USA, (October 20) 1998.

47. Heinzelman, W.; Chandrakasan, A.; Balakrishnan, H. Energy-Efficient Communication Protocol for Wireless Microsensor Networks. *Proceedings of the 33rd Hawaii International Conference on System Sciences (HICSS'00)*, January 2000.

CHAPTER 6

RESULTS AND DISCUSSION

6.1 INTRODUCTION

Virtual laboratories are a potential replacement for standard laboratory facilities. The use of these virtual resources can reduce cost and maintenance overheads for teaching institutions while still ensuring that students have access to real equipment. We will present some operational virtual laboratories around the world in Section 6.2.

Practical work (PW) in the laboratory to study wireless sensor networks (WSNs) is a very challenging task for the students in computer science and electronics. If we exclude systems which are dedicated to wire line networks such as those presented in,[36] no proposal of a virtual platform devoted to study the practical aspects of WSNs has been made up to now.

There are only expensive and cumbersome simulators such as OMNet++ (Objective Modular Network Testbed in C++) [OMN], Castalia,[6] Network Simulator_2 (NS2),[34] and Opnet,[26] for which PWs require the physical presence of the students as well as the availability of a great number of nodes in the classroom. For example, to make a simulation study on WSNs using NS2 environment, it is necessary to have skills in both Tool Command Language (Tcl) [33] and C++. Using virtual laboratories is a possible alternative to overcome these difficulties and contribute to make a PW on different aspects of WSNs such as clustering, routing, medium access control, aggregation, and security. There are several popular simulators for WSNs such as NS2, OMNet++, SensorSIM, GlomoSIM, QualNet, Opnet Modeler, Shawn, TOSSIM, EmStar, Avrora and so forth.

In what follows, we describe some of these simulators:

- Omnet ++
    - Website: Ref [25]
    - Platform: Microsoft Windows (with Cygwin), UNIX.
    - License: Free for academics and for nonprofit use.

The Objective Modular Network Testbed (OMNeT++) discrete event simulation environment is a tool used for the simulations of communication networks, multiprocessors, and various distributed systems. It is an open-source simulator based on C++ that was designed for the simulation of large systems and networks. A model in OMNeT++ consists of modules that communicate with each other using message passing.

- NS-2
  - Website: Ref [34]
  - Platform: UNIX (Linux, Solaris, Mac OS X uncertain)Microsoft Windows (with Cygwin)
  - License: Free

The network simulator (typically called NS-2, where the number 2 indicates the current version) is a widely used discrete event simulator targeted at networking research in general. It was written in a combination of C++ and an object-oriented dialect of Tcl, called OTcl. One reason for the popularity of NS-2 is its extensibility. Over time, many enhancements and extensions were developed, for example, to provide the support for wireless networks and mobile Ad hoc networks.

- SensorSIM
  - Website: Ref [31]
  - Platform: UNIX (Linux, Solaris, Mac OS X uncertain), Microsoft Windows
  - License: Free.

- GlomoSim
  - Website: Ref [19]
  - Platform: UNIX
  - License: Free for academics.

GloMoSim is a simulation tool based on the PARallel Simulation Environment for Complex systems (PARSEC) simulation environment. PARSEC is a C-based simulation language, which is used to represent a set of objects in the physical system as logical processes and interactions among these objects as time-stamped message exchanges.

GloMoSim supports a variety of models at different protocol layers, for example, CSMA and MACAW (MAC layer), flooding and DSR (network layer), and TCP and UDP (transport layer).

In addition, it supports different node mobility models, for example, the random waypoint model (i.e., a node chooses a random destination within the simulated area and moves toward this destination with a specified speed) and the random drunken model (i.e., a node periodically moves to a position chosen randomly from its immediate neighboring positions).

- QualNet
  - Website: Ref [30]
  - Platform: Microsoft Windows, Linux, Solaris
  - License: Commercial. Discounts are applied for research.

QualNet is a commercialized version of GlomoSim (presented above). It is produced by Scalable Network Technologies Inc.[15]

- Opnet Modeler
  - Website: Ref [26]
  - Platform: Microsoft Windows (NT, 2000, XP) and Solaris
  - Licence: Commercial, you can get it for free by registering in Opnet program for universities (time unknown).

We attempted to complete the theoretical study by implementing our own wireless sensor network virtual laboratory "Mercury".[10] It is based on an object-oriented design and a distributed approach, such as a self-organization mechanism, which is distributed at the level of each sensor. Knowing that the sensors are too expensive and not available, we developed "Mercury" to simulate the network partitioning into a number of clusters that are more homogeneous in a combination of metrics to produce a virtual topology. To determine and evaluate the results of the execution of the algorithms introduced above, the number of sensors (N) to deploy must be less or equal to 1000. There are two types of sensor node deployments in the sensor field, which are random and manual.

The design and development of a virtual laboratory platform (VLP) allowing learners to study and make PW on different aspects of mobile WSNs with lower costs by using UML methodology, allows the learners how to proceed in:

- Maintaining stable clustering structure and offering a better performance in terms of the number of re-affiliations using the proposed algorithm Energy Efficient and Safe Weighted Clustering Algorithm (ES-WCA).

- Detecting common routing problems and attacks in clustered WSNs, based on behavior level.
- Showing clearly the interest of the routing protocols in terms of energy saving processes and therefore, maximizing the lifetime of the global network.

Sensor nodes can be either thrown in mass or placed one by one in the sensor field. They can be deployed by:[1]

- Dropping from a plane
- Delivering in an artillery shell, rocket, or missile
- Throwing by a catapult (from a shipboard, for example)
- Placing in factory
- Placing one by one either by a human or a robot

Although the sheer number of sensors and their unattended deployment usually preclude placing them according to a carefully engineered deployment plan, the schemes for initial deployment must have following features:

- Reduce the installation cost
- Eliminate the need for any pre-organization and preplanning
- Increase the flexibility of arrangement, and
- Promote self-organization and fault tolerance.

6.1.1   SCENARIO

The number of sensors to deploy must be less than or equal to 1000 sensors, that is why we consider the following scenario, a helicopter that will drop the sensors in a forest or the border of a country for safety (cf. Fig. 6.1).

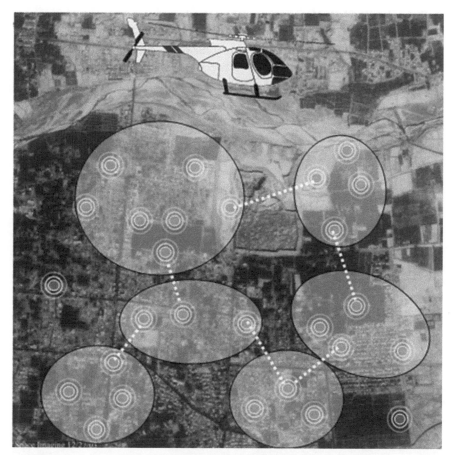

FIGURE 6.1   A helicopter drops the sensors randomly.
Source: Reprinted from Ref [29] with permission.

6.2   EXAMPLES OF VIRTUAL LABORATORIES

In what follows, we present some operational virtual laboratories around the world beginning with platforms made in our laboratory (LARESI), for advertising.

6.2.1 LARESI ROBOTIC PLATFORM

Nowadays, a remote manipulation of real machines like robots takes an important place in our daily life. In this context, we present in what follows the LARESI robotic platform,[4] offering learners in the field of robotics an interface to the possibility of remote controlling robot through the Internet (Fig. 6.2). Among the services offered by this platform, the learner can highlight the unpredictable nature of the communication channel, the Internet, by bringing the model of QoS control laws in the teleoperated system.

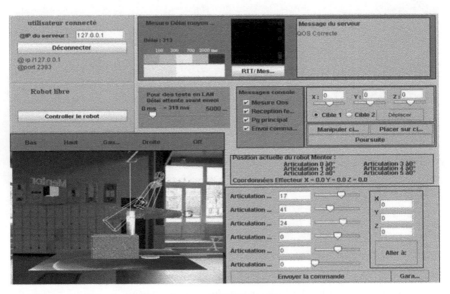

FIGURE 6.2   Robotic arm practical work (PW) interface.
Source: Ref [4] with permission.

Another functionality of this PW is the ability to perform a real object grasping task on a remote site based on the visual feedback from à stereoscopic system. The two cameras must be calibrated in order to extract the 3D coordinates of the target object. This operation is carried out entirely online, according to the following scheme (see Fig. 6.3).

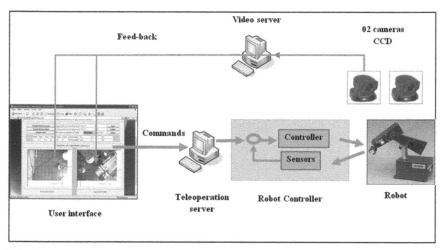

FIGURE 6.3   Diagram-based control vision.

6.2.2   BIOCHEMISTRY PRACTICAL WORK (PW)

Djamila Mechta et al.[23] used the specification of a biochemistry PW to demonstrate the potential use of our previous VLP. This platform has agent-based architecture as described in Figure 6.4.

The authors have chosen biochemistry specialty because almost all its courses require experimental work. The real experiments, in such a field, require expensive materials and might be dangerous for the students to manipulate.

The objective of this work is to design and implement the remote PW that we are interested in. They used multi-agent system engineering methodology to develop the system. This tool must be capable of creating a digital work environment where learners collaborate and cooperate among themselves to achieve the various stages of this PW.

The aim of this work consists in developing a specific virtual laboratory for a particular discipline and a specific PW based on our generic virtual laboratory.

The authors proposed a generic architecture based on agents which allow the learners to carry out distance PW and experiment with virtual devices (Fig. 6.5).

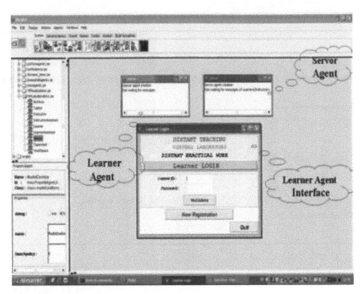

FIGURE 6.4   Principal interface of biochemistry PW.
Source: Ref [23] with permission.

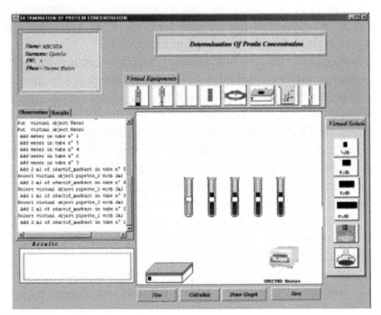

FIGURE 6.5   Experiment's workspace.
Source: Ref [23] with permission.

6.2.3 OPEN NETWORK LABORATORY (ONL)

Ken Wong et al.[36] created the Open Network Laboratory (ONL). It is an Internet-accessible virtual laboratory facility that can deliver a high-quality laboratory experience in advanced networking. Our experience with ONL indicates that it has the potential to improve student's understanding of the fundamental networking concepts and increase enthusiasm for the experimentation with complex technology. Furthermore, these benefits can be delivered with less effort from the instructor than using a traditional approach of socket programming and ns-2 simulation exercises. The system is built around a set of high performance, extendible routers that connect personal computers acting as end systems. The users configure their virtual network through the remote laboratory interface (RLI), an intuitive graphical interface.

The RLI's real-time charts and user data facility make it easy to directly view the effect of system parameters on traffic behavior. These features can enhance learning by providing users with multiple representations of network phenomena.

We describe how the ONL facilities have improved our ability to meet the instructional objectives and discuss some approaches to improving the laboratory experience.

6.2.4 DELTA'S VIRTUAL PHYSICS LABORATORY

Certainly, the most effective instrument to simplify and clarify the comprehension of any complex mathematical or scientific theory is through visualization. Moreover, using interactivity and 3D real-time representations, we can easily explore and hence, learn quickly in virtual environments. The concept of virtual and safe laboratories has vast potentials in education. With the aid of computer simulations and 3D visualizations, many dangerous or cumbersome experiments may be implemented in the virtual environments, with rather a small effort. Nonetheless, visualization alone is of little use if the respective simulation is not scientifically accurate.

Hence, a rigorous combination of precise computation as well as sophisticated visualization, presented through some intuitive user interface is required to realize a virtual laboratory for education. Sepideh Chakaveh et al.[7] introduced

Delta's Virtual Physics Laboratory, comprising a wide range of applications in the field of Physics and Astronomy, which can be implemented and used as an interactive learning tool on the World Wide Web (WWW).

6.2.5   REMOTE LABORATORY IN THE FIBER OPTICS AREA

J. John et al.[21] describes the details of the five experiments developed at IIT Kanpur for remote operation, as part of the virtual laboratories initiative of the MHRD, Government of India under the National Mission on Education through ICT (NMEICT). At present in India, there are no remote laboratories in the Fiber Optics area.

The laboratory used LabVIEW 8.6 as the software platform for remote access. It has been in operation for more than a year and is available for remote operation for about 4 h during a daytime. The details of the experiments, experience of running these laboratories and also its impact on teaching, and challenges are discussed.

Such laboratories would go a long way in meeting the acute shortage of specialized laboratories and trained faculty in India. In order to have a better impact on teaching, more such remote labs must be created with proper administrative and security features.

6.3   THE LABORATORY "MERCURY"

6.3.1   ARCHITECTURE OF "MERCURY"

The laboratory "Mercury" offers for learners a topic help (see Fig. 6.6). It contains the graphics used by the laboratory to make the PW more readable and understandable. On the other hand, a bit of simulators for WSNs such as TOSSIM[22] and Power-TOSSIM[32] are irrelevant with our goal and purpose and in order to avoid many complications, we established our own mercury simulator. When the user launches "Mercury," the following interface appears (see Fig. 6.7). In Fig. 6.8, we present the general architecture of our system. The different modules are classified according to their timing in execution.

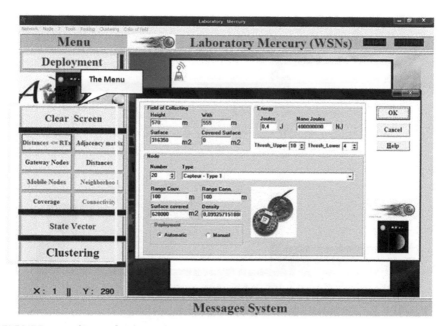

FIGURE 6.6 **(See color insert.)** Help topic of our virtual laboratory mercury.

6.3.2 A TOPIC HELP

The laboratory "Mercury" offers the possibility to the learners to select a type of sensors from five predefined types (see Table. 6.1). Each type has its characteristics (radius, energy, and so forth). The student can also introduce his/her own characteristics. The unity of the energy used is the nanojoule: 1 J = 10⁹ NJ.

6.3.3 DIFFERENT TYPES OF "MERCURY" SENSORS

TABLE 6.1 Different Types of "Mercury" Sensors

Type	Range (m)	Energy (J)
1	75	0.2
2	100	0.4
3	125	0.6
4	150	0.8
5	175	1

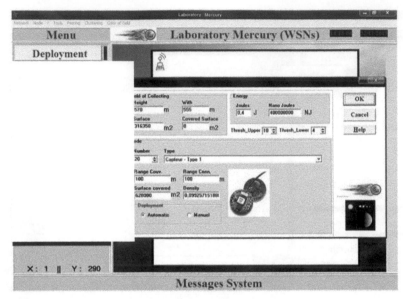

FIGURE 6.7 The virtual laboratory "Mercury" main interface.

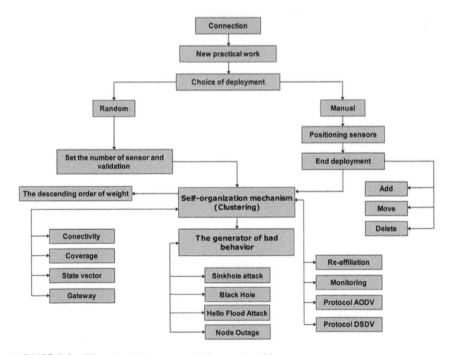

FIGURE 6.8 The virtual laboratory "Mercury" architecture.

6.4 ROUTING PROTOCOLS

6.4.1 AODV PROTOCOL

Being a reactive routing, protocol AODV uses traditional routing tables, one entry per destination and sequence numbers are used to determine whether the routing information is up-to-date and to prevent routing loops. Figure 6.11 will resume the whole process from route discovery until the route maintenance.[8] The maintenance of time-based states is an important feature of AODV which means that a routing entry which is not recently used expires. The neighbors are notified in case of route breakage. The discovery of the route from source to destination is based on query and reply cycles and intermediate nodes store the route information in the form of route table entries along the route. Control messages used for the discovery and breakage of route are as follows.[3]

6.4.1.1 ROUTE REQUEST (RREQ)

A route request (RREQ) packet is flooded through the network when a route is not available for the destination from the source. The parameters contained in the RREQ packet are presented below (cf. Fig. 6.9).

Source	Broadcast ID	Source sequence number	Destination	Destination sequence number	Hop count

FIGURE 6.9   Route request (RREQ) parameters.

Meanings of various parameters are:

- Source: The node which originated "route request".
- Destination: The destination for which a route is desired.
- Broadcast ID: A sequence number increments automatically when a node sends a new RREQ.
- Source sequence number: The current sequence number to be used for route entries pointing to (and generated by) the source of the RREQ.
- Destination sequence number: The last sequence number received in the past by the source for any route towards the destination.

- Hop count: The number of hops from the source to the node handling the request.

An RREQ is identified by the pair source and broadcast ID, each time when the source node sends a new RREQ and the broadcast ID is incremented. After receiving of the request message, each node checks the broadcast ID and source pair. The new RREQ is discarded if there is already a RREQ packet with the same pair of parameters.

6.4.1.2   ROUTE REPLY (RREP)

On having a valid route to the destination or if the node is a destination, a route reply (RREP) message is sent to the source by the node. The following parameters are contained in the RREP message (cf. Fig. 6.10):

Source	Destination	Destination sequence number	Hop count	Life time

FIGURE 6.10   Route reply (RREP) parameters.

Meanings of various parameters are:

- Source: The source node which issued the RREQ for which the route is supplied.
- Destination: The IP address of the destination for which a route is supplied.
- Destination sequence number: The destination sequence number associated with the route.
- Hop count: The number of hops from the Source IP Address to the destination IP Address.
- Lifetime: The time for which nodes receiving the RREP consider the route to be valid.

6.4.1.3   ROUTE ERROR (RERR)

The neighborhood nodes are monitored. When a route that is active is lost, the neighborhood nodes are notified by route error (RERR) message on both sides of the link. When the source receives this type of packets, it invalids the destination correspond status, stops the packets sending and place them in a queue. In the same time, it triggers a rerouting process.

6.4.1.3.1 HELLO Messages

The HELLO messages are broadcasted in order to know the neighborhood nodes. The neighborhood nodes are directly communicated. In AODV, HELLO messages are broadcasted in order to inform the neighbors about the activation of the link. These messages are not broadcasted because of a short time to live (TTL) with a value equal to one.

6.4.1.3.2 Path Discovery

When a source node does not have routing information about the destination, the process of the path discovery starts for a node with which source wants to communicate. The process is initiated by the source node by broadcasting of RREQ to its neighbors. Each neighbor either satisfies the RREQ by sending an RREP back to the source or rebroadcasts the RREQ to its own neighbors after increasing the hop count. If a node cannot satisfy the RREQ, it keeps track of the RREP parameters in order to implement the reverse path setup, as well as the forward path setup that will accompany the transmission of the eventual RREP.[18]

6.4.1.3.3 Reverse Path Setup

There are two sequence numbers (in addition to the broadcast ID) included in an RREQ such as the source and the last destination sequence numbers known to the source. The source sequence number is used to maintain freshness information about the reverse route to the source, and the destination sequence number specifies how fresh a route to the destination must be before it can be accepted by the source.[18] As the RREQ travels from a source to various destinations, it automatically sets up the reverse path from all nodes back to the source. To set up a reverse path, a node records the address of the neighbor from which it received the first copy of the RREQ. These reverse path route entries are maintained for at least enough time for the RREQ to traverse the network and produce a reply to the sender.[18]

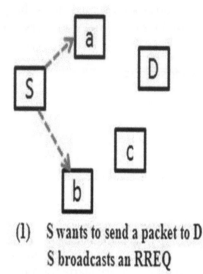

(1)   S wants to send a packet to D
      S broadcasts an RREQ

FIGURE 6.11   The whole process of AODV from route discovery until the route maintenance.
Source: Adapted from ref [5], with permission.

6.4.1.3.4   *Forward Path Setup*

Eventually, a RREQ will arrive at a node (possibly the destination itself) that can supply a route to the destination; a reverse path has been established to the source of the RREQ. As the RREP travels back the source, each node along the path sets up a forward pointer to the node from which the RREP came, records the latest destination sequence number for the requested destination. Nodes that are not along the path determined by the RREP, will time-out after active route time-out and will delete the reverse pointers. A node receiving an RREP propagates the first RREP for a given source node towards that source. If it receives further RREPs, it updates its routing information and propagates the RREP only if the RREP contains either a greater destination sequence number than the previous RREP or the same destination sequence number with a smaller hop count. The source node can begin data transmission as soon as the first RREP is received, and can later update its routing information if it learns of a better route.[18]

6.4.1.3.5 Route Table Management

In addition to the source and destination sequence numbers, other useful information is also stored in the route table entries and is called the soft-state associated with the entry. Associated with the reverse path routing entries is a timer, called the RREQ expiration timer. The purpose of this timer is to purge reverse path routing entries from those nodes that do not lie on the path from the source to the destination. The expiration time depends upon the size of the network. Another important parameter associated with routing entries is the route caching time-out, or the time after which the route is considered to be invalid.[8]

6.4.1.3.6 Path Maintenance

Failure of nodes not lying along an active path does not affect the routing to that path's destination. If the source node moves during an active session, it can reinitiate the route discovery procedure to establish a new route to the destination.

When either the destination or some intermediate node failure the risks to cut the route before the end of the current communication are very important.

Once the next hop becomes unreachable, the node upstream of the break propagates a RERR with a fresh sequence number (i.e., a sequence number that is one greater than the previously known sequence number) and hop count of "∞" to all active upstream neighbors. Those nodes subsequently relay that message to their active neighbors and so on. This process continues until all active source nodes are notified, it terminates because AODV maintains only loop-free routes and there are only a finite number of nodes in the network.[8]

There are two scenarios for the rerouting:

- Global rerouting: begins from the source node. This rerouting type is implemented in the most routing protocols, although it takes an important time and consumes lots of energy (cf. Fig. 6.12).
- Local rerouting: starts from the node where the failure takes place. This scenario is rapid and consumes less energy (cf. Fig. 6.13).

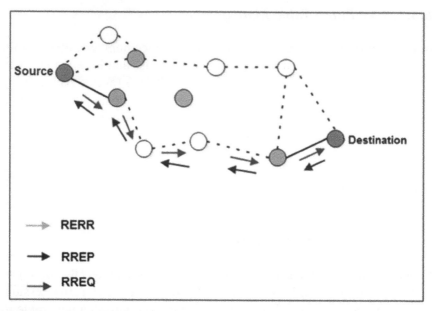

FIGURE 6.12   Global rerouting.
Source: Adapted from ref [5], with permission.

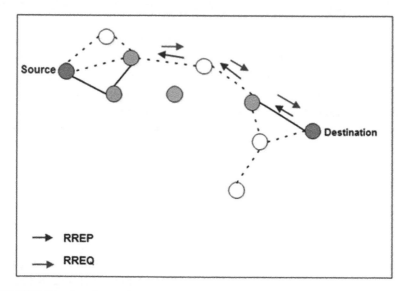

FIGURE 6.13   Local rerouting.
Source: Adapted from ref [5], with permission.

6.4.2 DSDV

DSDV is a table-driven protocol. Every node maintains a routing table that lists all available destinations, the number of hops to reach the destination and the sequence number assigned by the destination node. The sequence number is used to distinguish the stable routes from new ones and thus avoid the formation of loops. The nodes periodically transmit their routing tables to their immediate neighbors. When a route update with a higher sequence number is received, it replaces the old route. In case of different routes with the same sequence number, the route with the better metric is used. Updates have to be transmitted periodically or immediately when any significant topology change is detected.[28,35]

6.4.3 AODV CRITERIA

6.4.3.1 ADVANTAGES OF AODV

In what follows, we present the most significant advantages of AODV protocol:

- Because of its reactive nature, AODV can handle highly dynamic behavior networks.
- Unicast, Broadcast, and Multicast communication.
- On-demand route establishment with a small delay.
- Link breakages in active routes efficiently repaired.
- All routes are loop-free through use of sequence numbers.
- AODV reacts relatively quickly to the topological changes in the network and updating only the hosts that may be affected by the change.

6.4.3.2 LIMITATIONS OF AODV

We can enumerate the following limitations of AODV protocol:

- No reuse of routing info: AODV lacks an efficient route maintenance technique. The routing info is always obtained on demand, including for common cause traffic.
- AODV does not discover a route until a flow is initiated. This route discovery latency result can be high in large-scale mesh networks.

- AODV lacks support for high throughput routing metrics: AODV is designed to support the shortest hop count metric. This metric favors long, low bandwidth links over short, high bandwidth links.

6.4.4   DSDV CRITERIA

6.4.4.1   ADVANTAGES OF DSDV

In what follows, we present the most significant advantages of DSDV protocol:

- DSDV protocol guarantees loop-free paths.
- We can avoid extra traffic with incremental updates instead of full dump updates.
- Path selection: DSDV maintains only the best path instead of maintaining multiple paths to every destination. With this, the amount of space in the routing table is reduced.

6.4.4.2   LIMITATIONS OF DSDV

We can enumerate the following limitations of DSDV protocol:

- Wastage of bandwidth due to unnecessary advertising of routing information even if there is no change in the network topology.
- DSDV does not support the multi path routing.
- It is difficult to determine a time delay for the advertisement of routes.
- It is difficult to maintain the routing table's advertisement for a larger network.

Routing in sensor networks has attracted a lot of attention in the recent years[8,27] Table 6.2 shows the difference between reactive and proactive approach.

TABLE 6.2   Comparison of Proactive and Reactive Routing Protocols.

Proactive routing protocol	Reactive routing protocol
Reaction to the demand by diffusing requests	Control packets exchange
No routing table maintained continuously	Continuous updating of routing tables
Significant cost for the road's development	The roads are immediately available on demand
Considerable delay before sending a packet	The delay before sending a packet is minimal
No continuous traffic for unused routes	Control and update traffic maybe important and partially useless
Reduced use of the bandwidth	Large portion of the bandwidth to keep routing information up-to-date
Less energy consumption	More energy consumption for large networks

6.5   CHOICE OF NETWORK PARAMETERS

The choice of network parameters is done through tab "create a new network". A new window of new simulation will appear, these parameters are (see Fig. 6.14):

- The connectivity radius "R_T": Is the transmission range of sensor node. According to the literature, there is no difference between the radius of connectivity and coverage radius, the two terms are related to the same concept of radio range, that is to say, the maximum distance between a node with its neighbor's one-hop and with which it can communicate.
- The number of sensors to deploy: It must be less than or equal to 1000 sensors.
- The set up phase:

The cluster formation process means partitioning, virtually, the network by grouping all the sensors in disjoint groups called clusters and it has been widely studied in Ad hoc networks and wireless sensor networks. Figure 6.15 and 6.16 show an example of a random deployment of 200 sensors on a surface interest area of 570×555 m.

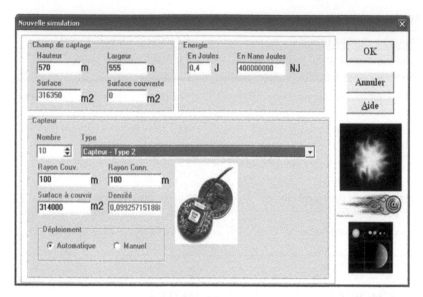

FIGURE 6.14   Choice of network parameters.

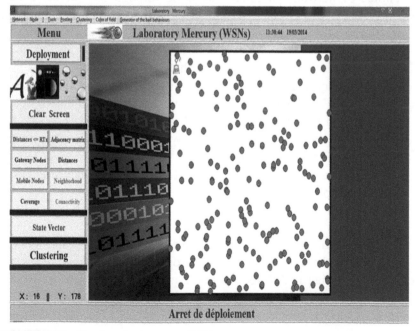

FIGURE 6.15   Random deployment of 200 sensors.

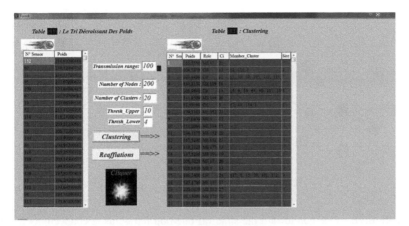

FIGURE 6.16   The descending order of weight of 200 sensors.

Flat routing protocols are quite effective in relatively small networks. However, they scale very bad to large and dense networks since, typically, all nodes are alive and generate more processing and bandwidth usage. In hierarchical-based routing, nodes play different roles in the network and typically are organized into clusters. Clustering (Fig 6.17) is the method by which sensor nodes in a network organize themselves into groups according to specific requirements or metrics. Each group or cluster has a leader referred to as cluster-head (CH) and other ordinary cluster members (CMs). The CH can be organized into further hierarchical levels.[16,37]

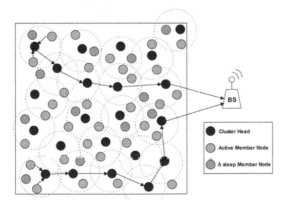

FIGURE 6.17   Cluster-based topology.

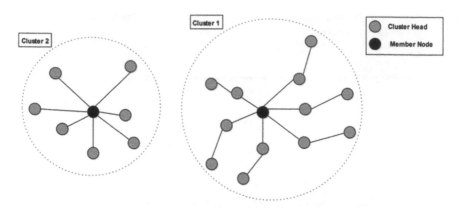

FIGURE 6.18 One-hop toward the sink, (a) one-hop intra-cluster connectivity and (b) multi-hop intra-cluster connectivity.

Most of the earlier works on clustering assume direct "one-hop" communication between member nodes and their respective CHs.[20,37] All the member nodes are at most two-hops away from each other (cf. Fig. 6.18(a)). One-hop clusters makes selection and propagation of CHs easy, however, multi-hop intra-cluster connectivity is sometimes required, in particular for limited radio ranges and large networks with limited CH count. Multi-hop routing within a cluster (cf. Fig. 6.18(b)) has already been proposed in wireless Ad hoc networks. More recent WSNs clustering algorithms allow multi-hop intra-cluster routing.[16,37]

Earlier cluster-based routing protocols such as LEACH[20] assume that the CHs have long communication ranges allowing direct connection between every CH and the sink (cf. Fig. 6.19). Although simple, this approach is not only inefficient in terms of energy consumption, it is based on unrealistic assumption. The sink is usually located far away from the sensing area and is often not directly reachable to all nodes due to the signal propagation problems.

A more realistic approach is multi-hop inter-cluster routing that had shown to be more energy efficient.[24,37] Sensed data are relayed from one CH to another until reaching the sink (Fig. 6.20).

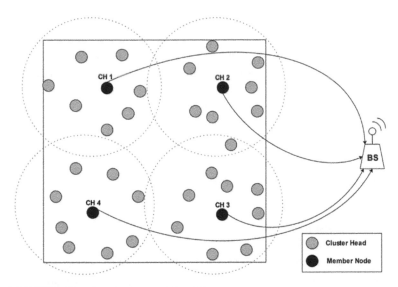

FIGURE 6.19   One-hop toward the sink.

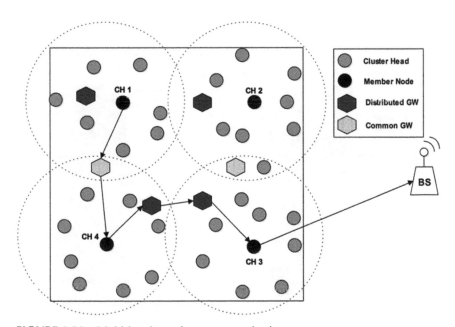

FIGURE 6.20   Multi-hop inter-cluster communication.

Direct communication between CHs is not always possible especially for large clusters (multi-hop clusters for instance). In this case, ordinary nodes located between two CHs could act as gateways (Gs) allowing the CHs to reach each other. A gateway node is either common or distributed. A common (ordinary) gateway is located within the transmission range of two CHs and thus, allows two-hop communication between these CHs. When two CHs do not have a common gateway, they can reach each other in at least three hops via two distributed gateways located in their respective clusters. A distributed gateway is only reachable by one CH and by another distributed gateway of the second CH cluster.

6.6 SIMULATION RESULTS

This section presents the implementation of the proposed approach using the Borland C++ language and the analysis of the obtained results. It is based on an object-oriented design and a distributed approach, such as a self-organization mechanism, which is distributed at the level of each sensor. Figure 6.21 and 6.22 show an example of cluster formation of 200 sensors and the state vector of each node in the network.

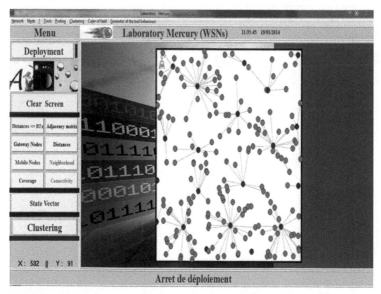

FIGURE 6.21 Network after clustering formation of 200 sensors.

N° Sensor	Position X	Position Y	Behaviour	Energy	Mobility	Connectivity	Distance	Weight	State	Role	Neighborhood	Packets_Sent	Packets_Received	Attack	Rate
1	130	342	1	49671	0	20	1311,548	165,255	Confident	ME 17 { 1; 15; 17; 3	74	46	..	0
2	0	456	1	49800	0	12	722,376	104,738	Confident	CH { 2; 32; 46; 5	56	27	..	0
3	32	235	1	49813	0	11	711,503	103,450	Confident	CH { 3; 59; 69; 1	53	24	..	0
4	420	548	1	49739	0	16	1008,132	134,113	Confident	ME 185 { 4; 11; 21; 3	69	64	..	0
5	61	44	1	49775	0	14	760,444	108,944	Confident	CH { 5; 9; 16; 26	83	36	..	0
6	388	211	1	49675	0	20	1077,514	141,851	Confident	ME 194 { 6; 14; 24; 2	82	32	..	0
7	534	12	1	49916	0	5	204,925	51,592	Confident	CH { 7; 44; 77; 1	29	6	..	0
8	270	372	1	49621	0	23	1448,193	179,519	Confident	ME 152 { 8; 12; 15; 2	107	38	..	0
9	16	63	1	49765	0	13	749,813	107,681	Confident	ME 5 { 5; 9; 16; 26	49	28	..	0
10	522	239	1	49740	0	15	781,064	111,206	Confident	ME { 10; 22; 24; 76		54	..	0
11	376	470	1	49531	0	28	1984,166	234,117	Confident	ME 152 { 4; 11; 12; 2	148	102	..	0
12	360	414	1	49637	0	21	1332,285	167,529	Confident	ME 152 { 8; 11; 12; 2	119	22	..	0
13	259	12	1	49799	0	13	870,130	119,713	Confident	ME 173 { 13; 18; 20; 66		35	..	0
14	411	219	1	49692	0	19	1036,296	137,530	Confident	ME 194 { 6; 14; 24; 2	90	75	..	0
15	196	401	1	49562	0	26	1734,217	208,722	Confident	ME 17 { 1; 8; 15; 17	89	42	..	0
16	31	75	1	49748	0	14	759,429	108,843	Confident	ME 5 { 5; 9; 16; 26	67	55	..	0
17	158	374	1	49592	0	24	1613,396	196,240	Confident	CH { 1; 15; 17; 1	87	91	..	0
18	230	59	1	49708	0	17	916,756	125,176	Confident	ME 137 { 13; 18; 20; 79		51	..	0
19	195	441	1	49559	0	27	1845,645	220,064	Confident	ME 152 { 15; 17; 19; 104		53	..	0
20	287	9	1	49853	0	10	660,554	98,155	Confident	ME 173 { 13; 18; 20; 38		10	..	0
21	388	466	1	49599	0	25	1647,351	199,835	Confident	ME 185 { 4; 11; 12; 2	142	29	..	0
22	489	292	1	49740	0	16	914,411	124,741	Confident	ME 194 { 10; 22; 24; 85		70	..	0
23	238	366	1	49690	0	23	1440,888	178,789	Confident	ME 136 { 8; 15; 17; 1	142	106	..	0
24	459	279	1	49691	0	19	1210,552	154,955	Confident	ME 194 { 6; 10; 14; 2	84	28	..	0
25	441	267	1	49640	0	21	1378,294	172,129	Confident	ME 194 { 6; 10; 14; 2	130	69	..	0

FIGURE 6.22 The state vector of 200 sensors.

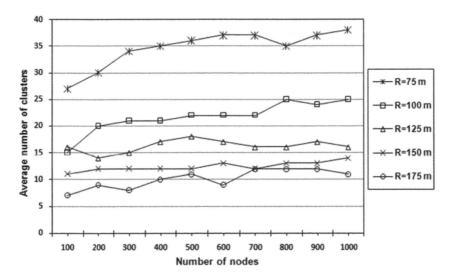

FIGURE 6.23 Average number of clusters versus number of nodes.

Figure 6.23 shows the variation of the average number of clusters with respect to the number of sensor nodes "N". The results are shown for varying transmission range (R) between 75 and 175 m. We observed that the number of clusters increase with the increase in the number of nodes. The experiments with a transmission range of 175 m give the best result, where the average number of clusters is very close to its value in the interval 400–500. The number of clusters remains stable in the interval of 700–900 and is equal to 12.[10,11,14] Figure 6.24 illustrates the variation of the average number of clusters, with respect to the transmission range. The results are shown for N, which varies between 200 and 1,000. We observe that the number of clusters decrease with the increase in the transmission range. This is due to the fact that a CH with a large transmission range will cover a large area.

Figure 6.25 depicts the average number of clusters that are formed with respect to the total number of nodes in the network. The communication range used in this experiment is 200 m. From the Figure 6.25, it is seen that ES-WCA consistently provides about 61.91% fewer clusters than DWCA and about 38.46% than SDCA, when there were 100 nodes in the network. When the node number is equal to 20 nodes, the performance of ES-WCA is similar to DWCA in terms of a number of clusters; however, if the node density had increased, ES-WCA would have produced constant-lyfewer clusters than SDCA and DWCA, respectively, regardless of the node number. Because of the use of a random deployment, the result of ES-WCA is unstable between 60 and 90.

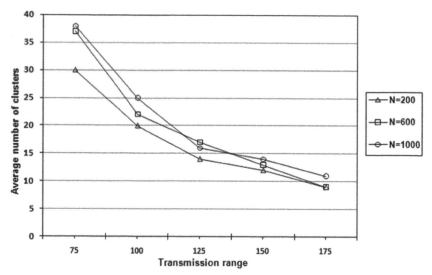

FIGURE 6.24   Average number of clusters versus transmission range (R).

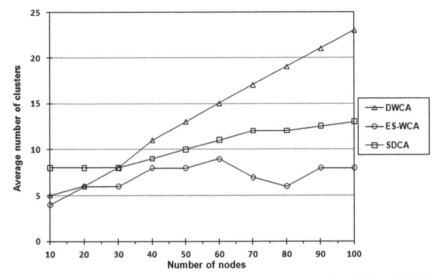

FIGURE 6.25 Average number of clusters versus number nodes (*N*) for ES-WCA, DWCA, and SDCA.

Therefore, the increase in the number of clusters depends on the increase of the distance between the nodes. As a result, our algorithm gave better performance in terms of the number of clusters, when the node density in the network is high, and this is due to the fact that ES-WCA generates a reduced number of balanced and homogeneous clusters, whose size lies between two thresholds such as $Thresh_{Upper}$ and $Thresh_{Lower}$ (re-affiliation phase) in order to minimize the energy consumption of the entire network and prolong sensors lifetime.

Figure 6.26 shows the variation of the average number of clusters, with respect to the transmission range. The results are shown for varying N. We observed that the average number of clusters decrease with the increase in the transmission range. As shown in Figure 6.26, the proposed algorithm produced 16–35% fewer clusters than WCA when the transmission range of nodes was 10 m. When the node density increased, ES-WCA constantly produced fewer clusters than WCA, regardless of the node number.

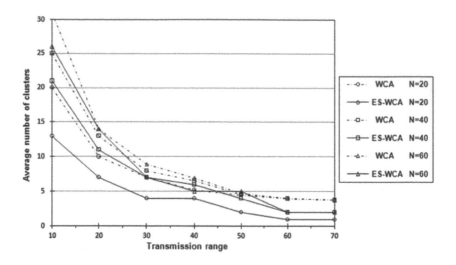

FIGURE 6.26   Average number of clusters versus transmission range ES-WCA and WCA.

With 70 nodes in the network, the proposed algorithm produced about 47–73% fewer clusters than WCA. The results show that our algorithm gave a better performance, in terms of the number of clusters, when the node density and transmission range in the network are high.

6.6.1   THE RE-AFFILIATION PHASE

We suggest re-affiliating the sensor nodes belonging to the clusters that have not attained the cluster size $Thresh_{Lower}$ to those that did not reach $Thresh_{Upper}$. Figure 6.27 interprets the average number of re-affiliations that are established with esteem to the total number of nodes in the network.[12,13] The number of re-affiliations incremented linearly when there were 30 or more nodes in the network for both WCA and DWCA. But for our algorithm, the number of re-affiliations increased starting from 50 nodes.

We submit to engender homogeneous clusters whose size is between two thresholds such as $Thresh_{Upper} = 18$ and $Thresh_{Lower} = 9$. According to the results, our algorithm presented a better performance in terms of the number of re-affiliations. The benefit of decreasing the number of re-affiliations mainly comes from the localized re-affiliation phase in our algorithm.

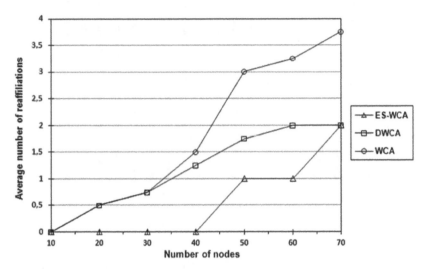

FIGURE 6.27   Average number of re-affiliations.

6.6.2   SHOW CLEARLY THE INTEREST OF THE ROUTING PROTOCOLS

The result of the remaining amount of energy per node for each protocol of AODV and DSDV is presented in Fig 6.28, such as R is equal to 35 m. As shown in the above-mentioned figure, the remaining energy for each node in the AODV protocol is greater than that in the DSDV protocol, such as AODV, which consumes 22.74% less than DSDV. According to the results, the network consumes 19.23% of the total energy when we used an AODV protocol (192,322,091 NJ). However, it consumes 41.97% with a DSDV protocol (419,740,129 NJ). We also observed that the network lost six nodes with DSDV, but only one node with AODV because of the depletion of its battery. This result clearly shows that AODV outperforms DSDV. This is due to the tremendous overhead incurred by DSDV when exchanging routing tables and because of the periodic exchange of the routing control packets. Consequently, our algorithm gave a better performance, in terms of saving energy when it is coupled with AODV.

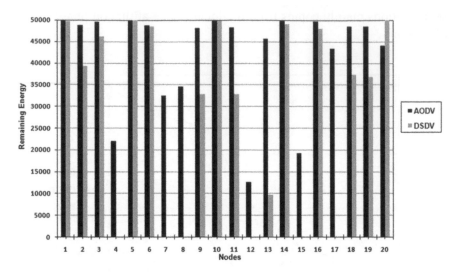

FIGURE 6.28   Remaining energy per node using ESWCA

In Figures 6.29–6.31, we evaluate the lifetime of the network by varying the number of nodes, such as R is equal to 70 m. We considered that the network will be invalid when the nodes in the neighborhood of the sink exhaust their energy, as illustrated in Fig 6.32 with the color red. There are nine nodes in an active state, but the network is invalid. We observed that the increase in the number of nodes does not have a significant impact on the lifetime of the network, except between N = 60 and N = 80. When there were 20 nodes in the network, the AODV increased the network duration by about 88.47% more than DSDV and about 57.9% for N = 100.

Also, this is due to the fact that in a DSDV protocol each node must have a global view of the network. This in turn increases the number of the exchanged control packets (overhead) in the whole network and it decreases the remaining energy of each node, which has a direct effect on the lifetime of the network.

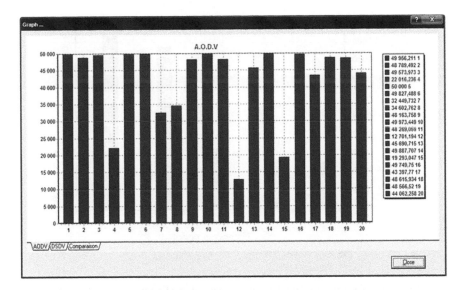

FIGURE 6.29   Remaining energy per node using protocol AODV.

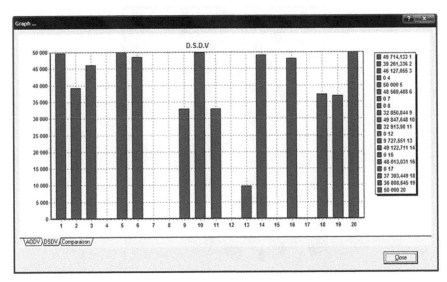

FIGURE 6.30   Remaining energy per node using protocol DSDV.

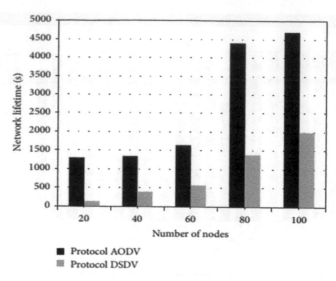

FIGURE 6.31   Network lifetime depending on number of nodes using ES-WCA.

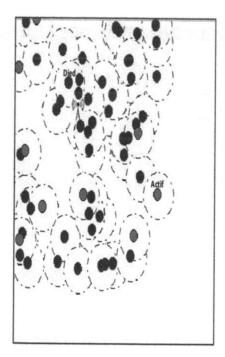

FIGURE 6.32   Snapshot showing the neighborhood of the sink exhausts their energy (N=60, R=30 m).

6.6.3 DETECTION OF COMMON ROUTING PROBLEMS AND ATTACKS IN CLUSTERED WSNS

6.6.3.1 SIMULATION USING 100 NODES

For the abnormal behavior experiment in the network,[10,11] we produced 100 sensors with 5 malicious nodes.

Figure 6.33 (b) shows the number of clusters formed according to the transmission range. Figure 6.34(a)–(c) shows the results of the experiment for a scenario with malicious nodes that are generated by the generator of bad behavior in Figure 6.35. The generated attacks are described in Chapter 3.

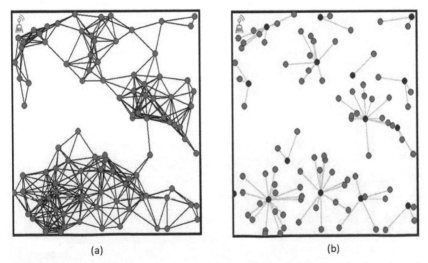

(a)   (b)

FIGURE 6.33   (a) Graph connectivity of 200 nodes and (b) network after clustering formation.

We can see that these nodes move from a normal state to an abnormal or suspicious state and finally, to a malicious state as expected. Table 6.3 shows the Ids of malicious nodes and their types of attacks during the distribution of a monitoring process in the network by the follow-up of the messages exchanged between the nodes. When Packets_sent (N1, N2), Packets_received (N3, N4).

Thus N1 is the number of packets sent before attacks and N2 is sent after attacks. While N3 is the number of packets received before attacks and N4 is received after attacks. We see that these malicious nodes increases by N1, as the sensors (16, 62), decrease by N1, like the sensor (3), increases by N3, as the sensor (41) and finally stop sending information like the node (94).

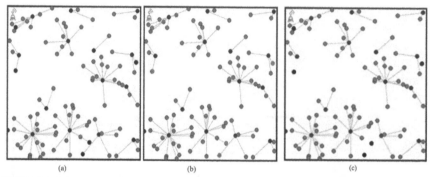

FIGURE 6.34   (a) Sensors with a blue color are abnormal but not malicious, (b) the grey sensors have a suspect behavior, and (c) the sensors with a black color are compromised and are exhibiting malicious behavior.

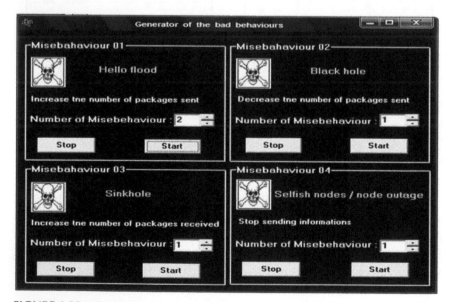

FIGURE 6.35   Generator of the bad behaviors.

TABLE 6.3   Detection of the Nature of Attacks (100 Nodes).

IDs	Packets_Sent	Packets_Received	Attack
16	(27 , 100)	(20 , 11)	Hello Flood
3	(13 , 1)	(5 , 8)	Black Hole
41	(5 , 4)	(4 , 76)	Sinck Hole
94	(5 , 4)	(4 , 4)	Node outage
62	(26 , 115)	(21 , 6)	Hello Flood

WSNs are considered in many cases as a stationary, but topology changes can be happened due to a weak mobility (new nodes join the network, existing nodes experience hardware failure or exhaust their batteries).[17] In other scenarios, the mobility can occur when nodes are carried by the external forces such as wind, water, or air[2] so that the network topology can be affected accordingly and changed slowly. This second kind of mobility, known as one form of strong mobility in literature in the sense, where nodes are forced to move physically in the deployment area, has been considered in this work.

In our previous paper,[9] the considered mobility has a particular sense by the fact that a mobile node does not move from one location to another in the space area of its own will, but in our case, it moves through the forces acting from the outside. These external forces can act from time to time sporadically. In contrary, the malicious node can use its own ability to move freely in the space area. The behavior of the malicious node by moving frequently inside the same cluster (case illustrated by Fig. 6.36) or from a cluster to another is a normal behavior to not attract the attention of the neighborhood and therefore be detected.

The idea of our algorithm to ensure the choice of a legitimate CH is to never elect a node that moves frequently and even it has the best performance metrics, but this malicious node does nothing just mobility, so in this new version of our algorithm (ES-WCA) detects the internal misbehavior of nodes during the distributed monitoring process in WSNs by the follow-up of the messages exchanged between the nodes.

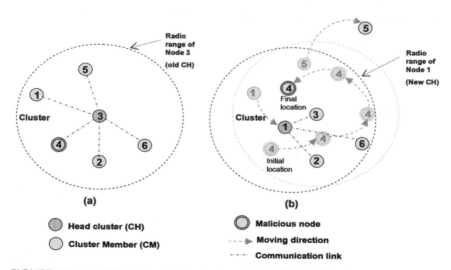

FIGURE 6.36 (a) Clustering mechanism in mobile WSNs before moving nodes and (b) after moving nodes 1, 5, and 4.

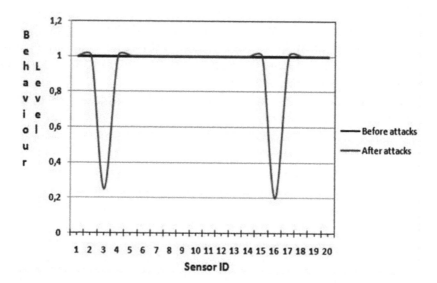

FIGURE 6.37 Behavior level of some sensors before and after attacks.

Mobility has an influence on energy and the behavior of sensors, for example, if the sensor moves 1 m away from its original location, its energy will diminish by 100,000 NJ and its behavior will also decrease by 0.001 units. This allows users to differentiate a malicious node (that moves frequently) from a legitimate node (that can changes position with reasonable distances). Since sensors nodes move because the forces acting from the outside, no power consumption for mobility must be taken into consideration in all simulations that we have carried for evaluation.

We note, from Figure 6.37, that the sensors nodes (3, 16) are malicious and have a behavior level that is less than 0.3. Their behavior decreased by 0.001 units when they move 1 m away from its original location.

6.6.3.2   SIMULATION USING 200 NODES

To illustrate the effect of an abnormal behavior in the network, in our experiments we propagated 200 nodes with five malicious nodes. The cases of the malicious nodes will pass from a normal node with a yellow color to an abnormal node with a blue color, to a suspicious node with a grey color, and lastly, to a malicious node with a black color.

All the cases of the CMs are discovered by their CH. Malicious CHs are disclosed by the base station.

Figure 6.38(b) displays the total of clusters established according to the transmission range. Figures 6.39(a–c) displays the measure results for a scenario with malicious nodes which are achieved by the generator of bad behavior.

The generated attacks are explained in Section 6.3. We can identify that these nodes migrate from a normal case to an abnormal or suspicious state and finally, to a malicious state as expected. Table 6.4 presents the Ids of malicious nodes and their categories of attacks in the course of the dissemination of a monitoring mechanism in the network by the follow-up of the messages exchanged between the nodes. When Packets_sent [$N1$, $N2$], Packets_received [$N3$, $N4$]. Thus, $N1$ is the total of packets sent before attacks, and $N2$ is sent after attacks, while $N3$ is the total of packets received before attacks and $N4$ is received after attacks. We regard that these malicious nodes increment $N1$, as the sensors (71,181), reduce $N1$, like the sensor (190), increment $N3$, as the sensor (162), and lastly break sending data like node (41). From Figure 6.40 it is observed that the sensor nodes (3, 17) are malicious and have a behavior level less than 0.3, its behavior decreased by 0.1 units, and when the monitor (CH) counts one failure an alarm is raised. However, packets from malicious nodes are not processed and no packet will be forwarded to them. The sensor node (11) has the behavior level less then threshold behavior so it will not be accepted as a CH candidate even if it has the other interesting characteristics (Eri, Ci, Di, and Mi). On the other side the behavior level in Figure 6.41 decreased by 0.001 units in our first work [DBK14] when the malicious node moves frequently.

We note that sensor (6) is suspicious so if it continues to move frequently its behavior will gradually be decreased until it reaches the malicious state. In this case, this node will be deleted from the neighborhood and finally it will be added to the blacklist.

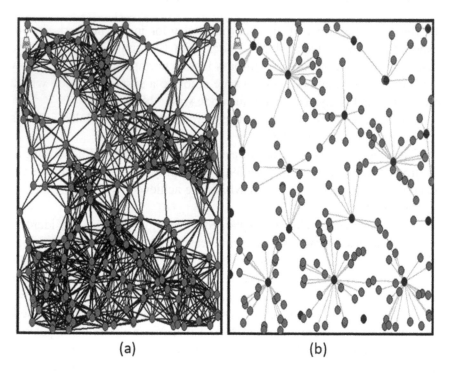

(a)   (b)

FIGURE 6.38   (a) Graph connectivity of 200 nodes and (b) network after clustering formation.

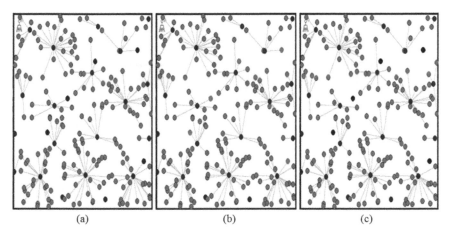

(a)   (b)   (c)

FIGURE 6.39 (a) Sensors with a blue color are abnormal but not malicious, (b) the grey sensors have a suspect behavior, and (c) the sensors with a black color are compromised and exhibit malicious behavior.

Therefore, we note that the network density has no effect on the detection of attacks that shows the performance of the monitoring phase (See Chapter 5).

TABLE 6.4 Detection of the Nature of Attacks (200 Nodes).

IDs	Packets_Sent	Packets_Received	Attack
41	(19 , 13)	(16 , 14)	Node Outage
71	(24 , 152)	(20 , 34)	Hello Flood
162	(15 , 8)	(22 , 112)	Sinck Hole
181	(16 , 179)	(26 , 42)	Hello Flood
190	(58 , 32)	(50 , 51)	Black Hole

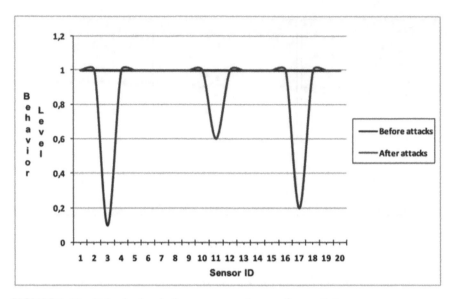

FIGURE 6.40   Behavior level of some sensors (moves frequently).

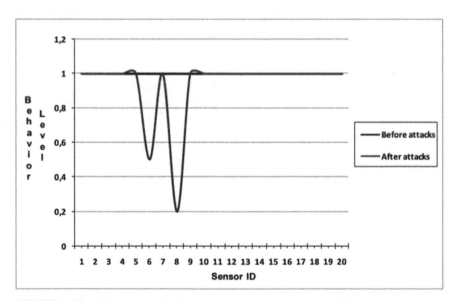

FIGURE 6.41   Behavior level of some sensors before and after attacks.

KEYWORDS

- practical work
- PARallel Simulation Environment for Complex systems
- virtual laboratory platform
- Open Network Laboratory
- remote laboratory interface

REFERENCES

1. Akyildiz, I. F.; Su, W. Sankarasubramaniam, Y.; Cayirci, E. Wireless Sensor Networks: A Survey. *Comput. Networks* **2002**, *38*(4), 393–422.
2. Ali, M.; Suleman, T.; Uzmi, Z. A. MMAC: A Mobility Adaptive, Collision-Free MAC Protocol for Wireless Sensor Networks. In *Proceedings of the 24th IEEE International Performance, Computing, and Communications Conference (IPCCC '05)*, IEEE, April, 2005, pp 401–407.
3. Ali Zakbar, A. (October) *Evaluation of AODV and DSR Routing Protocols of Wireless Sensor Networks for Monitoring Applications*; Master's Degree Thesis. Blekinge Institute of Technology: Sweden 2009.
4. Belhouari, A. Adaptation d'une station de télé-enseignement pratique au réseau sans qualité de service garantie. (Thèse de Magistère). 2010.
5. Beydoun, K. Conception d'un Protocol de Routage Hiérarchique pour les Réseaux de Capteurs, L'U.F.R des sciences et techniques de l'université de France COMTE, (Thèse de doctorat). 2009.
6. Castalia Official Site [Online], https://castalia.forge.nicta.com.au/index.php/en/index.html.
7. Chakaveh, S.; Zlender, U.; Skaley, D.; Fostiropoulos, J.; Breitschwerdt, D. DELTA's Virtual Physics Laboratory A Comprehensive Learning Platform on Physics and Astronomy. *IEEE*, **1999**, 421–423.
8. Dahane, A.; Berrached, N.; Hibi, A.; Loukil, A. Routing in Wireless Sensor Networks a Comparative Study: Between AODV and DSDV. International Conference on Embedded Systems in Telecommunications and Instrumentation (ICESTI'14), 2014.
9. Dahane, A.; Berrached, N.; Kechar, B. Energy Efficient and Safe Weighted Clustering Algorithm for Mobile Wireless Sensor Networks. *In The 9th International Conference on Future Networks and Communications (FNC'14)*, (Elsevier), August 17–20, Niagara Falls, Ontario, Canada. *Procedia Comput. Sci.* **2014**, *34*, 63–70,.
10. Dahane, A.; Berrached, N.; Loukil, A. A Virtual Laboratory to Practice Mobile Wireless Sensor Networks: A Case Study on Energy Efficient and Safe Weighted Clustering Algorithm. *J. Inf. Process Syst.* **2015**, *11*(2), 205–228, (June).

11. Dahane, A.; Loukil, A.; Kechar, B.; Berrached, N. Energy Efficient and Safe Weighted Clustering Algorithm for Mobile Wireless Sensor Networks. *Mobile Inf. Syst.* **2015**, 2015, 18.

12. Dahane, A.; Berrached, N.; Loukil, A. Balanced and Safe Weighted Clustering Algorithm for Mobile Wireless Sensor Networks. In *The 5th International Conference on Computer Science and its Applications (CIIA'15)*, Saida, Algeria, May 20–21, 2015; Chapter book, (Springer): Saida Algeria, **2015**; Vol. *456*, pp 429–441.

13. Dahane, A.; Berrached, N.; Loukil, A. Homogenous and Secure Weighted Clustering Algorithm for Mobile Wireless Sensor Networks. In *The 3rd International Conference on Control Engineering & Information Technology (CEIT '15)*, IEEE, Tlemcen, Algeria, May 25–27, 2015.

14. Dahane, A.; Berrached, N.; Loukil, A. Safety of Mobile Wireless Sensor Networks Based on Clustering Algorithm. *Int. J. Wireless Networks Broadband Technol. (IJWNBT)*. **2016**, *5*(1), 73–102, June.

15. Dargie, W.; Poellabauer, C. *Fundamentals of Wireless Sensor Networks: Theory and Practices*; Wiley Series on Wireless Communications and Mobile Computing, 2010.

16. Ding, P.; Holliday, J.; Celik, A. Distributed Energy Efficient Hierarchical Clustering for Wireless Sensor Networks. *IEEE International Conference on Distributed Computing in Sensor Systems (DCOSS'05)*, Marina Del Rey, CA, June, 2005.

17. Dong, Q.; Dargie, W. A Survey on Mobility and Mobility Aware Mac Protocols in Wireless Sensor Networks. *IEEE Commun. Surv. Tutorials* **2011**, *15*(1), 88–100.

18. Eperkins, C.; Mroyer, E. Ad-hoc On-Demand Distance Vector Routing, *Ad hoc networking*, Boston, USA, 2001, 173–219.

19. GlomoSIM Site [Online], http://pcl.cs.ucla.edu/projects/glomosim/.

20. Heinzelman, W.; Chandrakasan, A.; Balakrishnan, H. Energy-Efficient Communication Protocol for Wireless Microsensor Networks. *Proceedings of the 33rd Hawaii International Conference on System Sciences (HICSS'00)*, January, 2000.

21. John, J. *Remote Virtual Laboratory on Optical Device Characterization and Fiber Optic Systems: Experiences and Challenges. IEEE*, **2010**.

22. Levis, P.; Lee, N.; Welsh, M.; Culler, D. TOSSIM: Accurate and Scalable Simulation of Entire TinyOS Applications. *Proceedings of the 1st International Conference on Embedded Networked Sensor Systems (SenSys '03)*, November, 2003; 126–137.

23. Mechta, D.; Harous, S.; Djoudi, M.; Douar, A. A Collaborative Learning Environment for a Bioloy Practical Work. *Proceedings of the 12th International Conference on Information Integration and Web-based Applications & Services (iiWAS2010)*, Paris, 2010, 389–394.

24. Mhatre, V.; Rosenberg, C. Design Guidelines for Wireless Sensor Networks: Communication, Clustering and Aggregation. *Ad Hoc Networks* **2004**, *2*(1), 45–63, (January).

25. Omnet Official Site [Online], http://www.omnetpp.org.

26. OPNET Official Site [Online]. http://www.riverbed.com/products/performance-management control/opnet.html?redirect=opnet.

27. Parekh, A. K. Selecting Routers in Ad-Hoc Wireless Networks. *Proceedings of the SBT/IEEE International Telecommunications Symposium*, August, **1994**.

28. Perkins, C. E.; Bhagwat, P. Highly Dynamic Destination-Sequenced Distance-Vector Routing (Dsdv) for Mobile Computers. *ACM SIGCOMM Comput. Commun. Rev.* **1994**, *24*(4), 234–244.

29. Pham, C. *Data-Intensive Applications with WSNs*; Winter school on Wireless Sensor, CDTA, Algiers: Algeria, December 14th, 2014; pp 1–56.

30. QualNet Site [Online], http://www.scalable-networks.com/products/qualnet.php.

31. SensorSIM Site [Online], http://nesl.ee.ucla.edu/projects/sensorsim/.

32. Shnayder, V.; Hempstead, M.; Chen, B.-R.; Allen, G. W.; Welsh, M. Simulating the Power Consumption of Large-Scale Sensor Network Applications. *Proceedings of the 2nd International Conference on Embedded Networked Sensor Systems (SenSys '04)*, November, **2004**, 188–200.

33. Tcl Source Forge Project. http://tcl.sourceforge.net/.

34. The Network Simulator (ns-2). http://www.isi.edu/nsnam/ns.

35. Tripathi, K.; Agarwal, T.; Dixit, S. Performance of DSDV Protocol over Sensor Networks. *Int. J. Next-Gener. Networks (IJNGN)* **2010**, *2*(2), (June).

36. Wong, K.; Wolf, T.; Gorinsky, S.; Turner, J. Teaching Experiences with a Virtual Network Laboratory. *ACM SIGCSE Bull.* **2007**, *39*(1), 481–485.

37. Zeghilet, H. Le Routage dans les Réseaux de Capteurs Multimédia. Thèse de doctorat, **2013**.

INDEX